ISBN 978-1-330-46402-1
PIBN 10039756

1 MONTH OF
FREE
READING

at

www.ForgottenBooks.com

By purchasing this book you are eligible for one month membership to ForgottenBooks.com, giving you unlimited access to our entire collection of over 1,000,000 titles via our web site and mobile apps.

To claim your free month visit: www.forgottenbooks.com/free39756

English
Français
Deutsche
Italiano
Español
Português

www.forgottenbooks.com

Mythology Photography **Fiction**
Fishing Christianity **Art** Cooking
Essays Buddhism Freemasonry
Medicine **Biology** Music **Ancient**
Egypt Evolution Carpentry Physics
Dance Geology **Mathematics** Fitness
Shakespeare **Folklore** Yoga Marketing
Confidence Immortality Biographies
Poetry **Psychology** Witchcraft
Electronics Chemistry History **Law**
Accounting **Philosophy** Anthropology
Alchemy Drama Quantum Mechanics
Atheism Sexual Health **Ancient History**
Entrepreneurship Languages Sport
Paleontology Needlework Islam
Metaphysics Investment Archaeology
Parenting Statistics Criminology
Motivational

OUTLINES

OF

PHYSICAL GEOGRAPHY.

BY

GEORGE W. FITCH;

REVISED, WITH NOTES, ADDITIONS, AND AMENDMENTS, BY

ALPHONSO J. ROBINSON.

ILLUSTRATED WITH

𝔑𝔲𝔪𝔢𝔯𝔬𝔲𝔰 𝔐𝔞𝔭𝔰 𝔞𝔫𝔡 𝔈𝔫𝔤𝔯𝔞𝔟𝔦𝔫𝔤𝔰.

"Let me once understand the real geography of a country—its organic structure, if I may so call it; the form of its skeleton—that is, of its hills; the magnitude and course of its veins and arteries—that is, of its streams and rivers; let me conceive of it as a whole, made up of connected parts: and then the position of man's dwellings, viewed in reference to these parts, becomes at once easily remembered, and lively and intelligible besides."
DR. ARNOLD

CHANGED TO QUARTO FORM,

WITH QUESTIONS ON THE MAPS,

AND

AN ARTICLE ON THE PHYSICAL GEOGRAPHY OF THE UNITED STATES,

BY

CHARLES CARROLL MORGAN.

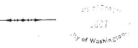

NEW YORK:

IVISON, PHINNEY, BLAKEMAN & CO., 47 AND 49 GREENE STREET.

CHICAGO: S. C. GRIGGS & CO., 39 AND 41 LAKE STREET.

1867.

DAVIES & KENT,
Stereotypers and Electrotypers,
183 WILLIAM ST., N. Y.

PREFACE.

THE following pages have been prepared with a view of supplying the want of a treatise on Physical Geography, adapted to the use of Schools and Academies. It is rather a remarkable fact, that among the multitude and variety of school-books prepared for the schools in the United States, there is not one devoted exclusively to this science. The consequence is, that Physical Geography, as a separate study, is very rarely taught in our schools, and that all, or most of the knowledge acquired respecting it, is what is incidentally obtained in pursuing other kindred studies.

The Author has aimed to present none but well-authenticated facts, and accordingly he has consulted the latest and most reliable authorities. Among the works from which valuable information has been obtained are "LYELL'S PRINCIPLES OF GEOLOGY," "MILNER'S GALLERY OF NATURE," "MILNER'S GEOGRAPHY," "KAEMTZ'S METEOROLOGY," "HUGHES' OUTLINES OF GEOGRAPHY," and "SOMERVILLE'S PHYSICAL GEOGRAPHY." To A. D. Bache, Superintendent of the United States Coast Survey, the Author is indebted for various Reports of the operations of that department, and for a Tide Table of the United States, which was specially prepared for this book.

The maps were compiled with the greatest care by Mr. George W. Colton, the aim being to exhibit the most remarkable and interesting features of Physical Geography, so far as they are capable of being represented to the eye.

It is proper to observe that, in the preparation of this treatise, no attempt at originality was made, but simply an effort to digest and arrange the more important facts in an intelligible style for learners. In many instances the Author has employed the phraseology of other writers, without always defacing the pages with quotation marks and references. Lengthened extracts, and those containing peculiar views of an author, are credited to their proper sources.

The Author can not refrain from expressing the hope that the book will meet the approbation of teachers, and excite in the minds of learners a desire for further attainments in this interesting department of science.

ADVERTISEMENT TO THE QUARTO EDITION.

IN order to accommodate in this work an increased number of maps (some of them on a large scale), and to admit the introduction of map questions on the same pages as the maps, or on the opposite pages, it has been changed to quarto form. In making this change, and in adding the map questions, however, no alteration has been made in the numbers of the paragraphs or descriptive Lessons. The references to the book, therefore, except for pages, will be found the same either in the duodecimo or quarto form.

It is believed that the above-mentioned additions much enhance the value of the work; since, by the more extensive use of maps, the subject is not only more perfectly illustrated, but a greater number of facts are taught through the medium of the eye, and thus are more clearly and durably impressed on the memory, while the student is oftener led to exercise his powers of observation and of philosophic deduction.

A carefully prepared chapter on the physical geography of the United States has been inserted in the Appendix, for the instruction of those who prefer seeking a knowledge of the subject in this connection. It embodies the results of the latest researches, and is one of the most complete essays on the natural character and resources of our country that has yet been published.

Many new pictorial illustrations, intended for instruction as well as embellishment, have also been inserted.

It is hoped the foregoing improvements will commend themselves to all who are interested in education, and will lead to a more extensive use of the book in the higher grades of schools.

CONTENTS.

CONTENTS.

APPENDIX.

LIST OF MAPS.

PHYSICAL GEOGRAPHY.

INTRODUCTION.

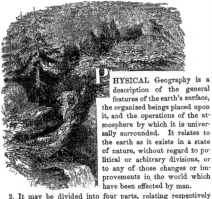

PHYSICAL Geography is a description of the general features of the earth's surface, the organized beings placed upon it, and the operations of the atmosphere by which it is universally surrounded. It relates to the earth as it exists in a state of nature, without regard to political or arbitrary divisions, or to any of those changes or improvements in the world which have been effected by man.

2. It may be divided into four parts, relating respectively to the following subjects: 1. The LAND, or solid portion of the earth's surface; 2. The WATER, or liquid portion; 3. The operations of the ATMOSPHERE; 4. ANIMAL and VEGETABLE life.

3. The *First Part* describes the extent and distribution of the land; the arrangement of the continents and islands; the magnitude and direction of the great mountain systems; and the situation and extent of the vast plains, upland and lowland, which constitute the most productive portions of the earth's surface. This part also relates to volcanoes, earthquakes, etc., in their relation to the character and aspect of the land portion of the earth.

4. The *Second Part* relates to the waters of the globe, whether salt or fresh; the origin, course, fall, and termination of rivers; the distribution and magnitude of lakes; and the extent, depth, tidal and other movements of the oceanic waters.

5. The *Third Part* treats of the operations of the atmosphere which surrounds our globe. It describes the laws which set the winds in motion, and the causes which influence their direction and velocity; it explains the phenomena of moisture, dew, rain, snow, and hail, and the various causes which are concerned in the regulation of climate.

6. The *Fourth Part* relates to organic existence, or the animals and plants distributed over the globe. It describes the great natural divisions of plants, the agencies which contribute to their diffusion, the food plants, and the regions where they are produced. It presents the orders of the animal kingdom, shows its diversity of organization, and its diffusion, and explains the zoological character of the different portions of the world.

7. The facts of Physical Geography are of a more permanent character than those which relate to the civil or political affairs of mankind. The boundaries of nations are frequently changed, either by conquests or treaties; new countries become peopled, and new states and territories organized; populous cities spring suddenly into existence; and the arts of civilization are rapidly carried to distant quarters of the earth. The varying condition of countries with respect to population, internal improvements, boundary lines, etc., is such as to require a constant correction of maps and statistical works, in order to make them correct exponents of political affairs.

8. Such fluctuations do not belong to the science of Physical Geography. The grand and majestic features which God has impressed on the face of our globe—its continents and oceans—its mountains, valleys, rivers, and lakes,—remain now, with all their prominent characteristics, very nearly the same as they have existed for centuries past.

9. It is true that natural agencies are at work, changing to a limited extent the face of nature. Volcanic action has rent the crust of the earth in numerous places, raising some portions and depressing others; some rivers have worked new channels, and formed extensive deltas at their mouths; and, as in Holland, vast areas of land have been rescued from the ocean by embankments and artificial modes of drainage. These and other alterations, considerable as they may appear, are comparatively unimportant as regards the world at large, and scarcely serve to qualify the remark, that the physical aspect of the earth has not greatly changed in modern times.

10. From what is here observed it must not be inferred that the earth has not been the scene of mighty convulsions. An examination of its surface shows that at very early periods most important changes successively took place. To explain those changes, and the causes which have led to the present state or condition of the earth belongs properly to the science of Geology.

11. Physical Geography explains many interesting facts of Civil Geography. It shows where nature has provided for the growth of cities, the peopling of states, the construction of railroads, canals, and other works of internal improvement; it points out what courses on the ocean the mariner must pursue in order to avail himself of its favoring winds and currents; and it explains what pursuits are best adapted to the people of different countries. The influence of mountains, rivers, seas, climate, and natural productions on the industry of people and the progress of nations is so great, that it is scarcely possible for one to possess a thorough knowledge of general geography without first understanding those facts which Physical Geography describes.

Questions.—1. What is Physical Geography? To what does it relate? 2. Into how many parts may it be divided? Name the subjects to which they relate. 3. What does the first part describe? To what does it also relate? 4. To what does the second part relate? 5. Of what does the third part treat? What does it describe? 6. To what does the fourth part relate? What does it describe?

Questions.—7. What is said of the facts of Physical Geography? Illustrate. Varying condition of countries? 8. Do such fluctuations belong to the science of Physical Geography? 9. How are certain changes of the earth's surface produced? What is observed of these alterations? 10. What does an examination of the earth's surface show? What belongs to the science of Geology? 11. Why is a knowledge of Physical Geography important?

PART I.

THE LAND.

LESSON I.
EXTENT AND DISTRIBUTION OF THE LAND.

HE surface of the earth consists of unequal portions of land and water. It has been estimated to contain about 196,500,000 English square miles. Of this area, the dry land is supposed to occupy about 51,000,000 square miles. Hence it will be seen the fluid portion predominates over the 'solid in the ratio of about 285 to 100. The extent of each division, however, can not be exactly ascertained, owing to the north and south polar regions not having been fully explored.

13. There is but little regularity in the arrangement of the land upon the globe. In some parts the coast is indented by deep bays and gulfs, in others the land projects into the ocean in capes and promontories, while the islands are scattered throughout the ocean, either singly or in irregular groups.

14. The distribution of the land is very unequal,—by far the greater portion being in the northern hemisphere. It has been calculated that there is about three times as much land north of the equator as south of it, and about two and a half times as much in the eastern as in the western hemisphere.

Its distribution through the different zones is as follows:

Northern Hemisphere.	Sq. Miles.	Southern Hemisphere.	Sq. Miles.
Arctic Zone	2,792,000	Antarctic Zone	Unknown.
Temperate Zone	24,488,000	Temperate Zone	3,305,000
Torrid Zone	9,949,000	Torrid Zone	10,466,000
Total	37,229,000	Total	13,771,000

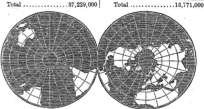

SOUTHERN HEMISPHERE.   NORTHERN HEMISPHERE.

15. The unequal distribution of the land may be most strikingly seen from an inspection of a map of the hemispheres, projected upon the plain of the horizon of London. The hemisphere, in which that city occupies the center, includes nearly all the land on the globe, while the other is almost covered with water. One may therefore be termed the *continental* or *land hemisphere*, and the other the *oceanic* or *water hemisphere*.

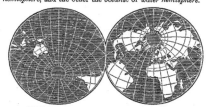

WATER HEMISPHERE.   LAND HEMISPHERE.

Note.—Before proceeding with the next lesson, the student will attend to the Map-Questions on the third page following.

LESSON II.
CONTINENTS.

16. THERE are two continents,—the Eastern Continent, or Old World, which includes Europe, Asia, and Africa, and the Western Continent, or New World, which includes North and South America. The Eastern Continent is styled the Old World, from its being the only one known to Europeans previously to the close of the fifteenth century. The terms Eastern and Western refer to the meridian of the Ferro Isles, from which longitude was formerly reckoned.

17. The Western Continent is about 8,700 miles in length from north to south. The greatest breadth of the continent is about 3,250 miles, and its least breadth, in the center, across the Isthmus of Panama, about 30 miles.

18. North America is the northern portion of the Western Continent. Its greatest length from north to south is about 5,600 miles, and its greatest breadth about 3,100 miles. It contains an area of about 7,493,000 square miles. The main body of the continent may be included within a triangle, whose base extends along the northern shores and whose vertical angle is in Mexico.

19. The eastern side of North America is penetrated by branches of the ocean, and consequently presents several peninsulas, and the western projects the long peninsula of California. These indented shores, which give to the continent a coast-line of 27,800 miles, or 1 mile to every 270 square miles of surface, are, with numerous rivers and lakes, the means by which civili-

Questions.—12. Of what does the surface of the earth consist? How many square miles does it contain? Square miles of the land? In what proportion does the fluid portion predominate over the solid? Why can not the extent of each division be exactly ascertained? 13. What is said of the arrangement of the land? 14. What is said of the distribution of the land? Which hemisphere contains the greater portion, the northern or southern? How much more land is there in the northern than in the southern hemisphere? In the eastern than in the western?

Questions.— 15. How may the unequal distribution of the land be most strikingly seen? Which hemisphere includes nearly all the land? How termed? 16. How many continents are there, and what countries do they respectively include? Why is the Eastern Continent styled the Old World? To what do the terms Eastern and Western refer? 17. Length of the Western Continent? Greatest breadth? Least breadth? 18. Length and breadth of North America? Area? General form of the continent? 19. What is the character of the coast-line? Extent of coast-line?

zation and commerce have been extended, and are now rapidly extending into the interior.

20. Among the numerous indentations of the Atlantic coast south of Labrador, are the Gulf of St. Lawrence, Bay of Fundy, Passamaquoddy Bay, Penobscot Bay, Massachusetts Bay, Cape Cod Bay, Buzzard Bay, Narraganset Bay, New York Bay, Raritan Bay, Delaware Bay, Chesapeake Bay, Albemarle Sound, Pamlico Sound, Gulf of Mexico, Bay of Honduras, etc.

21. South America is the smaller of the two divisions of the Western Continent. Its greatest length from north to south is about 4,600 miles, and its greatest breadth from east to west about 3,000 miles; its area is about 6,679,000 square miles. Its form is triangular. Its unbroken coast-line of 15,800 miles in extent, gives only a mile of sea-coast for every 423 square miles of surface, and presents few bays or even harbors.

22. The slow progress of civilization in South America has been attributed in a great measure to the want of bays and gulfs extending inland and affording maritime advantages to the interior regions. This disadvantage of unbroken coast-line is partly counterbalanced by the vast navigable streams of the Orinoco, Amazon, La Plata, and their branches.

LESSON III.

CONTINENTS (CONTINUED).

23. THE Eastern Continent is the largest mass of land upon the globe. It extends for about 10,000 miles from east to west, and about 7,800 from north to south. It contains an area of about 30,800,000 square miles, or about two and a sixth times as many as the Western Continent.

24. Europe is the smallest of the five grand divisions. Its greatest length from Cape St. Vincent, in the southwest, to the Gulf of Kara, in the northeast, is about 3,500 miles; its greatest breadth, from North Cape to Cape Matapan, is about 2,400 miles. The area of its surface, excluding the islands, amounts to about 3,506,000 square miles.

25. Europe is indented by numerous bays and seas on its

NAPLES AND VESUVIUS.
BAY OF NAPLES.

western and southern sides, in consequence of which the coast-line is of great extent, and in proportion larger than that of any other of the grand divisions. Its line of shores extends 20,000 miles; it therefore enjoys a mile of coast-line for every 175

square miles of surface, thus possessing great facilities for commercial enterprise. Europe is essentially the region of *peninsular* formations: it embraces the Scandinavian Peninsula (Norway and Sweden); the Peninsula of Denmark; the Peninsula of Spain and Portugal; the peninsulas of Italy and Greece.

26. Asia is the largest of the grand divisions of the earth. Its greatest length from east to west is about 5,600 miles, and its greatest breadth from north to south about 5,300 miles. It contains an area of about 15,909,000 square miles, or considerably more than is contained in both North and South America. It has a coast-line of 35,000 miles, and excluding the Arctic Ocean, which is scarcely navigable, there will be only 1 mile of sea-coast for every 454 miles of surface.

27. Africa, like South America, is a vast peninsula, being entirely surrounded by the waters of the ocean, except at the Isthmus of Suez, by which it is connected to Asia. Its greatest length from north to south is about 5,600 miles, and its greatest breadth from east to west about 4,700 miles. Its area is about 11,396,000 square miles. In consequence of its peculiar form, with no considerable peninsulas or sea indentations, its coast-line is only 16,000 miles, or only 1 mile of sea-coast for every 712 square miles of surface. On this account it is the most inaccessible, least civilized, and least known to civilized nations.

28. The following table exhibits the superficial extent of each grand division in English square miles, together with the length of coast-line possessed by each (in English miles); and the proportion which the latter of these measures bears to the former:

Grand Divisions.	Area.	Coast-line.	Sq. Miles of Area for One Mile of Coast.
North America	7,493,000	27,800	270
South America	6,679,000	15,800	423
Europe	3,506,000	20,000	175
Asia	15,909,000	35,000	454
Africa	11,396,000	16,000	712

GENERAL REMARKS ON THE CONTINENTS.

29. If we examine the map of the world, we may notice several features of similarity between the two continents. (1.) Each expands into broad extensive flats toward the north, while toward the south they narrow down to points, offering a rude resemblance to an inverted pyramid.

30. (2.) Both attain their greatest breadth about the parallel of 50° N., and are cut off by the ocean at about latitude 70°.

31. (3.) Each has a large portion of its area nearly detached; South America being joined to North America by the Isthmus of Panama, about 30 miles broad, and Africa being appended to Asia by the Isthmus of Suez, about 75 miles broad.

32. (4.) The peninsulas of both continents follow a southerly direction—as Scandinavia (embracing Sweden and Norway), Spain, Italy, Greece, Africa, Arabia, India, Malacca, Cambodia, Corea, and Kamtchatka in the one, and South America, California, Florida, and Alaska in the other. There are few important exceptions to this rule; as the Peninsula of Yucatan, in Central America, and of Denmark, in Europe, which project toward the north.

Questions.—20. Principal indentations of the Atlantic coast south of Labrador? 21. What is said of South America? Greatest length? Breadth? Area? Form? Extent of coast-line? 22. Slow progress of civilization in South America? What compensation is there for its unbroken coast-line? 23. What is said of the Eastern Continent? Its length and breadth? Area? 24. What is the comparative size of Europe? Its length and breadth? Area? 25. What is said of its coast indentation? Extent of sea-coast?

Questions.—What is said of its peninsular character? What large peninsulas does it embrace? 26. Comparative extent of Asia? Its length and breadth? Area? Coast-line? 27. What is said of Africa? Its length and breadth? Area? Extent of coast-line? 29. What may be noticed by examining a map of the world? What is the first feature of similarity mentioned? 30. What is the second? 31. What the third? 32. The fourth? What exceptions are there to the fourth remark?

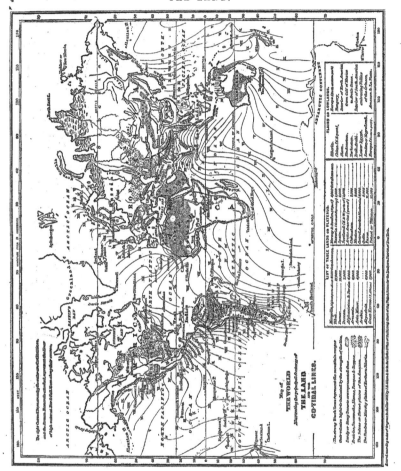

Map of
THE WORLD
Illustrating the principal features of
THE LAND
and the
CO-TIDAL LINES.

QUESTIONS ON THE MAP.

EXTENT AND OUTLINE OF THE CONTINENTS.

Which of the two continents is the largest? Which tends to the north? How many degrees nearer than the other to the earth's pole does it lie? Which tends farthest south? How many miles is it than the other to the south? Ans. About 19. Which, then, has the greatest range of title, and how many degrees difference is there? How many miles from north to south is the Eastern continent than the Western? Ans. Nd far from 900. How many miles more than the Western does it extend in a due east and west line? Ans. Nearly 500. What three of the grand divisions of land this has a triangle in the? What great island in the India has a ... shape? How are the two southern grand divisions ... with the adjoining seas of the continent? Ans. They are, 6th, east peninsulas. In their character? Ans. What does the ... with does five ... to the hm... gular seas were fixed to ... this in the same direction? Does it ... pear to be a general rule that the lands widen toward the north and grow narrower toward the ... ? Do the ... ern and ... shores of the great triangular seas, as we ... advance northward, ... only side from a north and ... line? Ans. Yes, with this divergence often found to average?

Ans. 23½° from a north and south line, or nearly the area as the ... tion of the earth's axis to a perpendicular to the ... of its orbit. Are the ... ern points of the great triangular ... masses generally levelled and mountainous, or ...? On each side of Africa is ... ere a deep inward bend? Is there a ... far ... bel on the ... side of South America? Of ... alia? Of the ... Falls of Hindoostan? In ... you far from the ... ? ... fled to in the ... not two ... ighs ... hat the ... coss of ... great bodies of ... are merely ... tal, or in, in same degree, the result of like causes operating in accordance with fixed laws? What two grand divisions are the most ... regular in outline? What two ... and in large peninsulas, and ... ... ere ... ake exceedingly varied ... ? Has the ... tern or western side of America the more ... uried ... tour, and ... fore the longer shore line?

RELIEFS OF THE CONTINENTS.

On which side of the Western Continent are the principal mountain-chains? Nearest what ocean are they situated, or toward which does the continent present its shortest and most abrupt slope? On which side of the Eastern Continent are the chief mountain-chains? Nearest what two oceans do they lie, or toward which is the short slope of this continent? Around what ocean, then, may it be said the principal mountain-chains of the earth are arranged, and the short slopes of the continent inclined? Which and where is the highest mountain-chain on the globe?

INSULAR REGIONS.

Between what two oceans do we find the ... of large islands on the globe? With what grand division do ... out of ... ? ... ... near to be ... ciated? What, where, are they ... cies called? Ans. The ... The ... genus of what great ... ocean are ... with an immense ... uber of small ... nsist of ... of the ... est archipelago appear to be wholly ... What has its ... ence here—together with other facts—led any ... men to conclude? Ans. That it ... pies the area of a once unbroken ... which has been ... merged, and whose lofty ... mits form the basis of ... of the ... insular groups and ... lia.

---

33. The great point of dissimilarity between the Eastern Continent and the Western is in the prevailing direction of the land, which extends from east to west in the former, and from north to south in the latter.

34. Comparing the western shores of Europe and Africa with the eastern shores of North and South America, a mutual adaptation to unite may be observed in the advancing and retreating shape of the land. Thus the great convexity of Western Africa is opposite to the indentation of the Gulf of Mexico, and the convexity of the Brazilian shore is opposite to the Gulf of Guinea. The idea has been entertained, from this peculiar outline, that the two continents once formed an undivided territory, which some great convulsion separated.

LESSON IV.

ISLANDS.

35. ISLANDS differ vastly in size, some being miniature continents, with systems of mountains, rivers, and lakes, while others are mere banks of sand or points of rocks just raised above the level of the waves. The largest island in the world is Australia; it is 2,400 miles from east to west, 1,700 miles from north to south, and contains an area of about 3,000,000 square miles. Its extent of coast-line is about 8,000 miles.

36. The following table exhibits the area of some of the largest islands, and their relative size, as compared with the area of the State of New York (46,220 square miles).

Name.	Area in Square Miles.	Comparative Size.
STATE OF NEW YORK	46,220	1.00
Iceland	30,000	.65
Ireland	32,515	.70
Newfoundland	36,000	.78
Cuba	43,000	.90
Great Britain, including England, Wales, and Scotland	88,827	1.82
Sumatra	120,000	2.59
Papua, or New Guinea	200,000	4 33
Madagascar	200,000	4.33
Borneo	270,000	5.08
Australia	3,000,000	64.81

37. Of the small islands, the most remarkable is Rockall, in the North Atlantic; it is only a hundred yards in circumference, and is situated 260 miles from the north coast of Ireland, and 180 from any other land.

38. Islands occur under various conditions, in chains, clusters, archipelagoes, or singly.

39. The principal chains are adjacent to some main shore, and on this account they are sometimes termed continental islands. They are long in proportion to their breadth, and follow each other in succession along the margin of the continents. America offers numerous examples of this kind of islands. On the northwestern coast there is a long chain of them, beginning with Vancouver's Island on the south. Another range

Questions.—33. What is the great point of dissimilarity between the Eastern and Western continents? 34. What may be observed by comparing the western shores of Europe and Asia with the eastern shores of North and South America? 35. How do islands differ? The largest island and its extent? 36. What is the area of Iceland, and what its relative size as compared with the State of New York? Ireland? Newfoundland? Cuba? Great Britain? Sumatra? Papua, or New Guinea? Madagascar? Borneo? Australia? 37. Give particulars of the island of Rockall. 38. Under what various conditions do islands occur? 39. Where are the principal chains? Shape and arrangement? What chains belonging to America are mentioned? Other instances?

occurs at the southern extremity of South America, extending from Chiloe to Cape Horn. · To this class also belong the Aleutian Isles, which form a chain between North America and Asia, in the North Pacific, and the Kurile and Japan Isles, stretching along the eastern Asiatic coast.

40. *Clusters*, sometimes called oceanic islands, are those which occur at a distance from continents. They are very numerous in the Pacific and Indian oceans. They usually contain one or two principal members centrally situated with reference to others of smaller size, as, for example, the Marquesas and Society groups.

41. An *archipelago* is a sea interspersed with numerous islands. The term archipelago was originally applied to those islands which lie between the shores of Greece and Asia Minor. The principal archipelagoes are the Caribbean, or Antilles, in the West Indies ; the Maldive and Laccadive, in the Indian Ocean ; the Dangerous, Louisiade, and Great Cyclades, in the Pacific Ocean.

42. *Single islands* at a great distance from any other shore are of rare occurrence. St. Helena, remarkable for being the

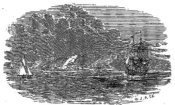

ST. HELENA.

place of Napoleon's last exile, is 1,800 miles from the coast of Brazil, 1,200 from the coast of Africa, and 680 from Ascension Island, the nearest point of land. · Ascension Island is also 520 miles from its next neighbor, the Isle of St. Matthew.

43. A vast number of islands are volcanic. Some are at present the scenes of fiery convulsions. Volcanic islands are found principally in the Indian and Pacific oceans, though some occur in high northern and southern latitudes. They are characterized by a considerable elevation, with a precipitous coast.

44. In the Grecian Archipelago, the Old Kaimeni, a small islet, was thrown up somewhat more than two centuries before the Christian era. A second appeared in the year 1573, called the Little Kaimeni, and a third was formed in the year 1707, called the New Kaimeni.

45. In the year 1811, the temporary island of Sabrina rose off the coast of St. Michael, one of the Azores. It attained the height of 300 feet, was about a mile in circumference, but gradually subsided, and wholly disappeared by the close of February, 1812. In 1813 there were five hundred feet of water at the spot.

Questions.—40. What are clusters ? Where numerous ? How usually arranged ? Examples ? 41. What is an archipelago ? How originally applied ? Principal archipelagoes ? 42. What is said of single islands ? St. Helena ? Ascension Island ? 43. What is said of volcanic islands ? Where principally found ? How characterized ? 44. What volcanic islands were formed in the Grecian Archipelago ? 45. Give the particulars of the formation and disappearance of the island of Sabrina. 46. Of Graham Island. 47. Islands and reefs in the Indian and Pacific oceans ? To what owing ?

GRAHAM ISLAND.

46. The most recent instance of an island formed by volcanic action was Graham Island, which rose in the Mediterranean Sea, southwest of Sicily, in July, 1831. A column of water was seen rising from the sea like a water-spout, followed by dense steam, and an island which gained the height of 200 feet, and a circumference of three miles. Toward the close of the year, this island gradually sank beneath the waves, forming a dangerous shoal.

LESSON V.
CORAL ISLANDS.

47. A vast number of islands and reefs* in the Pacific and Indian oceans are of coral formation. They owe their existence to the work of countless myriads of the coral-polyps, which inhabit those seas, and which flourish only in the warmer regions of the globe.

48. Coralline structures are sometimes of enormous extent. On the northeast coast of Australia is a reef of coral called the Great Barrier Reef, having a length of nearly 1,000 miles, and being in one part unbroken for a distance of 350 miles. Some groups of coral islands in the Pacific are from 1,100 to 1,200 miles in length, by 300 or 400 in breadth, as the Dangerous and Radack archipelagoes, for example. The Maldive Islands, situated in the Indian Ocean, forming a chain of 470 geographical miles, are composed throughout of a series of circular assemblages of islets, all formed of coral.

49. The following description of coral animals and their operations is from Hughes' "Manual of Geography :" "The coral reefs of the Pacific, as well as those in other parts of the globe, are all produced by the secretions of the coral insect, and the process by which they are formed is one of the most curious and instructive phenomena which the natural world presents to view. The architects of these wonderful structures are *polyps* of minute size, and of various species, but all possessing a general similarity of form and structure. They consist, to appearance, of a little oblong bag of jelly, closed at one end, but having the other extremity open, and surrounded by tentacles (usually six or eight in number), set like the rays of a star.

50. "Multitudes of these tiny creatures are associated in the secretion of a common stony skeleton, that is, the coral, or madrepore, in the minute orifices of which they reside, protruding their mouths and tentacles when under the water ; but the moment they are molested, or become exposed to the atmosphere, withdrawing by sudden contraction into their holes. It is proved by observation that these creatures are unable to exist at a greater depth than twenty or thirty fathoms ; so that the numberless coral islands of the Pacific, and other seas, must be based upon submarine rocks or mountains, though it was at one time supposed that they were raised, by the process described above, from the bottom of the sea."

* REEF, a chain or range of rocks lying at or near the surface of the water.

51. Coral formations are of four different kinds, namely, atolls, or lagoon islands, encircling reefs, barrier-reefs, and coral fringes.

52. An *atoll* consists of a circular strip or ring of coral surrounding a shallow lake or lagoon in its center. The circular reefs just raise themselves above the level of the sea, with an average breadth of a quarter of a mile, oftener less, and are surrounded by a deep and often unfathomable ocean. The annexed cut represents one of these circular islands inclosing a lagoon of tranquil water.

CORAL ISLAND.

The usual form of such islands may be seen in the section below.

SECTION OF A CORAL ISLAND.

a, a, habitable part of the island, consisting of a strip of coral, inclosing a lagoon. *b, b,* the lagoon.

53. Lagoons are found in a very large proportion of the coral islands. They were found in twenty-nine out of the thirty-two islands visited by Beechey in his voyage to the Pacific. The largest was thirty miles in diameter, and the smallest less than a mile. There is almost always a deep narrow passage opening into the lagoon, generally on the leeward side, which is kept open by the efflux of the sea, as the tide goes down, and through this channel ships may sail into the inclosed waters and find a good harbor.

54. *Encircling-reefs* are those which extend around mountainous islands, commonly at a distance of two or three miles from the shore, rising on the outside from a very deep ocean, and separated from the land by a channel 200 or 300 feet deep. The Caroline Archipelago exhibits examples of this structure. Otaheite (Tahiti), the largest of the Society group, is an instance of an encircled island of the most beautiful kind, being hemmed in from the ocean by a coral band, at a distance varying from half a mile to three miles.

55. *Barrier-reefs* are similar in their structure to the two preceding classes, but differ from them in their position with regard to the land. The largest of this class is the Great Barrier Reef off the northeast coast of Australia, before alluded to (48). It rises up in the ocean at an average distance of from

20 to 30 miles from the shore, and extends to the distance of about 1,000 miles.

56. The action of the waves as they dash upon this reef has been admirably described: "The long ocean-swell being suddenly impeded by this barrier, lifted itself in one great continuous ridge of deep blue water, which, curling over, fell on the edge of the reef in an unbroken cataract of dazzling white foam. Each line of breaker ran often one or two miles in length with not a perceptible gap in its continuity. There was a simple grandeur and display of power and beauty in this scene that rose even to sublimity. The unbroken roar of the surf, with its regular pulsation of thunder, as each succeeding swell fell first on the outer edge of the reef, was almost deafening, yet so deep-toned as not to interfere with the slightest nearer and sharper sound. . . . . . But the sound and sight were such as to impress the spectator with the consciousness of standing in the presence of an overwhelming majesty and power."

57. The Florida reefs are of this class. By examining a map of the waters south of Florida, it will be seen that they are studded with a range of islands called the Florida Keys. These keys rise but a few feet, perhaps from six to eight or ten, or at the utmost to twelve or thirteen feet above the level of the sea. They begin to the north of Cape Florida, and extend in a southwesterly direction, gradually receding from the land opposite Cape Sable. Farther to the west they project in a more westerly course as far as the Tortugas Islands, which form the most western group. Most of these islands are small, the largest of them, such as Key West and Key Largo, not exceeding ten or fifteen miles in length; others only two or three, and many scarcely a mile. Their width varies from a quarter to a third or half of a mile, the largest barely measuring a mile across.

58. The reef extends parallel to the main range of keys, for a few miles south or southeast of it, following the same curve, and never receding many miles from it. The distance be-

BARRIER-REEF.

tween the reef and the main range of keys varies from six to two or three miles. Between this reef and the main range of keys there is a broad, navigable channel, extending the whole length of the reef, varying in depth from eighteen to forty feet.

Questions.—51. Kinds of coral formations ? 52. The atoll ? What does it surround ? Height of the circular reef ? Breadth ? How surrounded ? 53. Lagoons ? Lagoons found by Beechey ? Their extent ? Openings into lagoons ? 54. Encircling-reefs ? Example ? Otaheite ? 55. Barrier-reefs ? The largest of this class ?

Questions.—Describe the appearance of this reef. 57. To what class do the Florida reefs belong ? Island south of Florida ? The height ? Where do they begin and where extend ? Their size ? 58. Where does the reef extend ? Distance between the reef and the main range of keys ? Channel ?

59. The great danger of this reef arises from the fact that throughout its whole range it does not reach the surface of the sea, except in a few points, where it comes almost to the level of low-water mark. It therefore presents a range of most dangerous shoal grounds, upon which thousands of vessels, as well as millions of property, have been wrecked.

60. *Coral-fringes* are those formations which extend along the margin of a shore, and have no lagoons.

61. Captain Basil Hall, in his "Voyage to Loo-Choo," makes the following observations on coral islands: "The examination of a coral reef during the different stages of one tide is particularly interesting. When the sea has left it for some time, it becomes dry, and appears to be a compact rock exceedingly hard and ragged.; but no sooner does the tide rise again, and the waves begin to wash over it, than millions of coral worms protrude themselves from holes on the surface which were before quite invisible. These animals are of a great variety of shapes and sizes, and in such prodigious numbers, that in a short time the whole surface of the rock appears to be alive and in motion.

62. "The most common of the worms at Loo-Choo (an island in the Pacific east of China), was in the form of a star, with arms from four to six inches long, which it moved about with a rapid motion in all directions, probably in search of food. Others were so sluggish that they were often mistaken for pieces of the rock; these were generally of a dark color, and from four to five inches long and two or three round. When the rock was broken from a spot near the level of high water, it was found to be a hard, solid stone; but if any part of it were detached at a level to which the tide reached every day, it was discovered to be full of worms, all of different lengths and colors, some being as fine as thread and several feet long, generally of a very bright yellow, and sometimes of a blue color; while others resembled snails, and some were not unlike lobsters and prawns in shape, but not above two inches long.

63. "The growth of coral ceases when the worm which creates it is no longer exposed to the washing of the tide. Thus a reef rises in the form of a gigantic cauliflower, till its top has gained the level of the highest tides, above which the worm has no power to carry its operations, and the reef, consequently, no longer extends itself upward. The surrounding parts, however, advance in succession till they reach the surface, where they also must stop. Thus, as the level of the highest tide is the eventual limit to every part of the reef, a horizontal field comes to be formed coincident with that plane, and perpendicular on all sides. The reef, however, continually increases, and being prevented from going higher, must extend itself laterally in all directions; and this growth being probably as rapid at the upper edge as it is lower down, the steepness of the face of the reef is preserved; and it is this circumstance which renders this species of rock so dangerous to navigation. In the first place, they are seldom seen above the water; and in the next, their sides are so abrupt that a ship's bows may strike against the rock before any change of soundings indicates the approach of danger.

64. "For a long time it was supposed that the coral formations were raised from the floor of the fathomless ocean by the unaided efforts of these little creatures, but more accurate observations have proved that the animals cease to live at a greater depth than twenty or thirty fathoms. . . . . As some of these islands are elevated 200 and 300 feet above the sea-level, it is evident that they must have been raised by submarine forces; in short, that the volcano and the earthquake must have been employed in rearing them to their present elevation. Mr. Darwin has traced those regions throughout the Pacific, in which upheaval and depression alternately prevail. Thus a band of *atolls* and encircled islands, including the Dangerous and Society archipelagoes, constitutes an area of subsidence more than 4,000 miles long and 600 broad. To the westward, the chain of *fringing-reefs*, embracing the islands of the New Hebrides, Solomon, and New Ireland, form an area of elevated coral. Farther westward, another area of subsidence is met with, including the islands of New Caledonia, and the Australian barrier."

LESSON VI.
MOUNTAINS.

CLIMBING THE ALPS.

65. MOUNTAINS are the most considerable elevations of the surface of the earth. They are of various heights, the loftiest having an elevation of more than five miles above the level of the sea. Though generally sterile, and unsuited for the residence of man, they have their uses in the economy of nature. They accumulate the moisture of the clouds, and feed the rivers which water and fertilize the plains below. They increase the surface of the earth, and consequently its productions. To their gigantic proportions, their lofty projections, and their broken and varied forms, are we largely indebted for sublime and savage, or beautiful and picturesque scenery.

66. There are but few insulated mountains, or mountains remote from other masses, and ascending abruptly from a level country. The examples are chiefly volcanic, as Mount Egmont, in New Zealand, and the Peak of Teneriffe, on one of the Canary Islands. The usual arrangement is in groups or chains, the members of which are connected at the base. The term *system* is applied to a series of chains, groups, and parallel ranges lying in the same general direction, though detached. The highest points are usually about the middle of the range.

67. The great mountain systems of the two continents follow the prevailing direction of the land in each; those of the Western World running north and south; those of the Eastern, east and west. The course of secondary chains, as the Apennines in Italy, the Dovre-field in Norway, and the Ghauts in India, corresponds with the greatest length of those peninsulas.

68. The highest known mountain on the globe is Mount Everest or Gahoorishanka, in Asia. It belongs to the Himalaya range, and is situated in about longitude 86° 50′ east. Its summit is 29,002 feet above the level of the sea. In the same range, to the east, is Kunchinjinga, the next in height (28,178 feet), and which was, until recently, considered the highest

Questions.—59. Danger of this reef? Vessels and property destroyed? 60. Coral-fringes? 65. What are mountains? What is said of their heights? Uses of mountains? 66. Insulated mountains? Examples? What is the usual arrangement of mountains? To what is the term *system* applied?

Questions.—67. Direction of the great mountain systems of the two continents? Course of the secondary chains? 68. The highest known mountain, its situation and height? Kunchinjinga? Name and height of the loftiest mountain in South America? In North America? In Africa? In Europe?

mountain on the globe. According to recent statements the loftiest mountains known in the other four grand divisions are, in South America, Tupungato, one of the Chilean Andes (22,450 feet); in North America, Popocatapetl, in the volcanic chain of southern Mexico (17,884 feet); in Africa, Mount Kenia (supposed about 20,000 feet); in Europe, Mont Blanc (15,760 feet).

69. The mountains of the torrid zone are capable of being inhabited by man to a very considerable height. Under the equator the line of perpetual snow is not less than about 16,000 feet above the level of the sea. As we approach toward the poles this line gradually descends, rendering the mountains of the temperate zones inhabitable to no very considerable elevation. On Mount Blanc the snow-line is about 8,500 feet above the

MOUNT CHIMBORAZO.

sea-level, and at the height of 6,000 feet the climate is of very great severity.

70. Table of the heights above the sea of some remarkable inhabited sites.

	Feet.		Feet.
Geneva, Switzerland	1,450	Hospital of the Great St. Bernard, Alps	8,170
Madrid, Spain	2,170	Santa Fé de Bogota, capital of New Granada	8,650
Jerusalem, Mount Zion	2,200		
Priory of Chamouni, Switzerland	3,346	Chuquisaca, capital of Bolivia	9,250
Palace of the Escurial, Spain	3,520	Pass of Santa Maria, Alps, highest permanent habitation in Europe	9,272
Teheran, Persia	3,785		
Ispahan, "	4,140	Quito, capital of Equador	9,540
Great Salt Lake City, U. S.	4,300	Ladak, city, Little Tibet	9,995
Hampelbaude, highest inhabited house of Prussia	4,300	Cuzco, ancient capital of Peru	11,380
Splugen, village, Switzerland	4,711	La Paz, city, Bolivia	12,226
Mont Louis, Eastern Pyrenees, highest town of France	5,171	Puno, city, Peru	12,870
Cabool, Afghanistan	6,380	Villages on south side of the Himalayas	18,000
Post-house on Mont Cenis, Alps	6,453	Potosi, Bolivia, highest city of the globe	13,350
Soglio, village in the Grisons, highest village in Europe	6,714	Antisana, shepherds' huts, Equador	13,354
Hospital of St. Gothard, Alps	6,808	Tacora, village, Peru	13,690
Mexico, city	7,570	Rumihausi, post-house, Peru	15,540
Arequipa, city, Peru	7,852	Ancomara, "	15,724

71. The summits of the loftiest mountains have never been reached, though some adventurous travelers have attained heights where man can find no local habitation. The difficulties encountered in ascending elevated mountains arise from the precipitous character of the surface, the vast accumulation of snow, the intense cold, and the rarity or thinness of the atmosphere. Travelers have found the rarefied atmosphere on high mountains to cause a bleeding from the nose and eyes, and to produce other unpleasant effects.

72. In the following list are given some remarkable heights which have been reached:

Questions.—69. Habitation of mountains in the torrid zone? The line of perpetual snow under the equator? Toward the poles? 70. Height of Geneva? Madrid? Jerusalem? Great Salt Lake City? Mexico? Santa Fé de Bogota? Chuquisaca? Quito? Potosi? Shepherds' huts, Equador? The post-house, Rumihausi, Peru? 71. What is said of heights attained by adventurous travelers? The difficulty of ascending elevated mountains? 72. Name the heights reached on the following mountains, and by whom: Mount Blanc, Jungfrau, Ortler Spitz, Peak of Demavend, Ararat, Pamir, Peter Botte, Mouna-Kaah, Mount Egmont, Silla de Caraccas, Pichincha, Chimborazo.

Sites.	Names.	Dates.	Heights.
Mount Blanc, Alps	Dr. Paccard and James Balma	Aug., 1786	15,750
Jungfrau, "	The brothers Meyer, of Arau	— 1811	12,872
Ortler Spitz, "	Three peasants of the Tyrol	1804	12,850
Peak of Demavend	Mr. Taylor Thompson, 1st European	Sept 9, 1837	14,700
Ararat	Professor Parrot, and five attendants	Oct. 9, 1829	17,210
Pamir, Central Asia	Lieutenant John Wood	Feb. 19, 1835	15,600
Peter Botte, Mauritius	Captain Lloyd and officers	Sept 7, 1832	2,500
Mouna-Kaah, Owhyhee	Mr. David Douglas	Jan., 1834	13,587
Mount Egmont, New Zealand	Dr. Dieffenbach	Dec., 1840	8,839
Silla de Caraccas	Humboldt and Bonpland	Jan., 1800	8,688
Pichincha, Andes	Bouguer and Condamine	— 1786	15,924
Chimborazo, "	A. von Humboldt	June 23, 1802	19,286
Purgeool, Himalaya	Captain A. Gerard	Oct., 1818	19,411
Chimborazo, Andes, point reached, highest point of the globe ever attained by man.	M. Boussingault and Colonel Hall	— 1831	19,699

LESSON VII.

THE MOUNTAIN SYSTEMS OF THE WESTERN CONTINENT.

73. NORTH AMERICAN MOUNTAINS.—North America contains three great mountain systems,—the Rocky Mountains, the mountains of the West Coast, and the Apalachian system. It embraces besides, the elevated regions of the Ozark Mountains, the highlands of Labrador and the Arctic coast, and the mountain traversed plateaus of Mexico and Central America.

74. The *Rocky Mountains* constitute the most extensive mountain system of North America. They extend from north to south through all the wider parts of the continent; or from the shores of the Arctic Ocean on the north to about the parallel of 32° on the south.

ROCKY MOUNTAIN SCENERY.

The northern portion is divided into numerous ranges, with hardly more elevation than from one to two thousand feet. As they advance southward their height increases, and many of their summits rise above the snow-line. They attain the most considerable elevations between the 55th and 38th parallels. The average heights between these limits is from seven to eight thousand feet. The highest known summits of the system are *Mount Brown* (15,690 feet) and *Mount Hooker* (15,700 feet), both near the line of the 52d parallel.

75. Numerous passes occur in the range of the Rocky Mountains, the most noted of which is that known as the South Pass, near the 41st parallel. It is at an altitude of more than 7,000 feet above the level of the sea, and affords a passage so easy of access that a wagon drawn by horses might travel through it. Thousands of emigrants, with their cattle, every year traverse this pass on their way to the valleys of the Pacific.

76. The *Mountains of the West Coast* extend along the Pacific, from Cape St. Lucas to the Peninsula of Alaska. They

Questions.—73. What three great mountain systems does North America contain? What other elevated regions does it embrace? 74. What is said of the Rocky Mountains? Where do they extend? What is said of the northern portion? Where do they attain the most considerable elevation? What is the average height between these limits? Which are the highest summits of the system? 75. Where is the principal pass situated? What is said of it? 76. Where do the mountains of the West Coast extend? What minor ranges do they embrace? What peaks, their height and situation? What is said of the Sierra Nevada? Where do the gold regions of California lie?

embrace the *Sierra Nevada* of California and the *Cascade Range* of Oregon. Among the elevated peaks are *Mount Shasta*, in the Cascade Range (14,390 feet); *Mount St. Elias*, near the 60th parallel (17,900 feet); and *Mount Fairweather*, near the 59th parallel (14,750 feet). The Sierra Nevada rises above the snow-line, and attains a main elevation of between seven and eight thousand feet. To the west of this range lie the gold regions of California.

77. The *Apalachian* or *Alleghany* ranges constitute the third great mountain system of North America. They extend along the eastern side of the continent, and within the older settled portion of the United States. They terminate at the south near the 34th parallel, and at the north near the shores of the Gulf of St. Lawrence. This range is broken by the valley of the Hudson River and Lake Champlain. The southern portion (to which the term "Alleghany" is alone applied) consists of numerous parallel ridges separated by longitudinal valleys. The northeastern section embraces the *Green Mountains* of Vermont and the *White Mountains* of New Hampshire. The *Adirondack Mountains*, which extend through the northern part of New York to the west of Lake Champlain, form a part of this system. Intermediate between the Adirondack Mountains and the main range of the Alleghanies, lie the *Catskill Mountains*, which are terminated on the north by the valley of the Mohawk, and on the east by the Hudson River.

WHITE MOUNTAINS.

78. The Apalachian range extends a distance of 1,500 miles, with an average height of from 2,500 to 3,000 feet. Among the highest elevations are, *Black Dome* or *Mitchell's High Peak*, in North Carolina, the highest of the entire system (6,707 feet); *Mount Washington* of the White Mountains

(6,288 feet); *Mount Marcy* of the Adirondack range, the highest in New York (5,379 feet); and *Mansfield Mountain*, the highest of the Green Mountains (4,430 feet).

79. The *Ozark Mountains* are about 300 miles in length, and extend from the State of Missouri through the northeast part of Arkansas into the Indian Territory. They vary from one to two thousand feet in height. The highlands of Labrador and the Arctic coast have a very broken and rugged surface, and an average elevation of from one to two thousand feet. They contain a great number of lakes, and possess a climate of very great severity.

80. The principal mountains in Mexico are isolated peaks, many of which are active volcanoes. Several of these peaks lie along the line of the 19th parallel: among them are *Colima*, *Jorullo*, *Toluca*, *Popocatepetl*, and *Orizaba*.

81. SOUTH AMERICAN MOUNTAINS.—South America likewise contains three mountain systems,—the chain of the Andes, the mountains of Guiana, and the mountains of Brazil.

82. The *Andes*, or Cordilleras de los Andes (*Chains of the Andes*), commence on the north near the Isthmus of Panama, and run in a southerly direction to the Straits of Magellan. In the south of Chile and in Patagonia they form the coast-line, and at the greatest distance, in about the middle of Chile, are but 100 miles from the sea. Their general breadth rarely exceeds from 200 to 250 miles; but between the 20th and 25th parallels they are upward of 400 miles across.

83. The Andes are divided, according to the countries through which they extend, into the Columbian, Peruvian, Bolivian, Chilean, and Patagonian Andes.

84. The *Columbian Andes* begin at the commencement of the mountainous region on the north, and extend to the 4th degree of south latitude. They have an average height of from 11,000 to 12,000 feet, and the highest peaks exceed 20,000 feet. The most elevated of these is *Chimborazo* (21,424 feet), which was long supposed to be the loftiest mountain in the New World.

85. The *Peruvian* and *Bolivian* Andes extend from the 4th to the 28th parallel of south latitude, and are remarkable for the great number of elevated summits they contain, several of which have an altitude of more than 20,000 feet. Many of the passes in this portion of the Andes are between 15,000 and 16,000 feet in height.

86. The *Chilian Andes* are remarkable for containing, according to recent reports, the highest known summit in the Western Continent—*Tupungato*, which attains an altitude of 22,450 feet above the level of the sea. The *Patagonian Andes* rise abruptly from the shores of the Pacific, which they border to a distance of about 1,000 miles from Cape Horn to the 40th parallel of south latitude. The average height of the Patagonian Andes is from 2,000 to 3,000 feet, though in some places they attain an elevation of 9,000 feet.

Questions.—77. What is said of the Apalachian or Alleghany range? Where do they extend? Where do they terminate at the south? At the north? By what valley is this range broken? What is said of the southern portion? What mountains does the northeastern section embrace? What is said of the Adirondack Mountains? The Catskill Mountains? 78. What distance does the Apalachian range extend? What is the average height? Which are among the highest summits? Give the height of each. 79. The Ozark Mountains? What is their height? Describe the highlands of Labrador and the Arctic coast. 80. What is said of the principal mountains in Mexico? What volcanic peaks lie along the line of the 19th parallel?

Questions.—81. How many mountain systems does South America contain, and how are they designated? 82. Describe the situation of the Andes. Where do they form the coast-line, and where are they at the greatest distance from the coast? What is their general breadth? Their greatest breadth? 83. How are the Andes divided? 84. Where are the Columbian Andes situated? Average height? Highest peaks? Chimborazo? 85. Between what parallels do the Peruvian and Bolivian Andes extend? For what are they remarkable? How high are many of the passes in this portion of the Andes? 86. For what are the Chilean Andes remarkable? Tupungato? What is said of the Patagonian Andes? Average height?

87. The *Mountains of Guiana* extend from the river Orinoco in a southeasterly direction nearly to the mouth of the Amazon. The most western of these mountains is distinguished as the *Parime Mountains*, and the eastern as the *Sierra Acáray*. *Mount Maravaca*, the highest summit of the system, has an elevation of about 11,000 feet.

88. The *Mountains of Brazil* embrace a great extent of country; they lie mostly in narrow chains or ridges, the most elevated summits of which appear to be less than 6,000 feet high.

LESSON VIII.
MOUNTAIN SYSTEMS OF THE EASTERN CONTINENT.

89. EUROPEAN MOUNTAINS.—The continental part of Europe embraces two important mountain systems,—one in the south and the other in the north, the former being by far the most extensive. The Ural and Caucasus mountains, though usually classed among those of Europe, form natural boundaries between Europe and Asia, and therefore belong as much to one division as the other. The south mountain region includes the Balkan, the Alps, the Carpathian Mountains, the mountains of the Spanish peninsula, and the Apennines of Italy. The north mountain regions extend through Norway and Sweden, constituting what are sometimes called the Scandinavian Mountains.

90. The *Balkan Mountains* are situated south of the river Danube, and extend from the shores of the Black Sea, in a westerly direction, through the central part of Turkey in Europe. A branch leaves the main chain about the meridian of 23° east, and stretches northward to the banks of the Danube; this may be called the North Balkan Mountains. South of the Balkan are three considerable branches,—the Little Balkan, the Despoto Dagh, and the chain of Mount Pindus, the latter stretching through the whole of the Grecian peninsula. The mountains of the Balkan system have generally but a moderate elevation, not exceeding on the average from 2,000 to 3,000 feet above the sea-level. In some instances, however, they reach as high as 8,000 or 9,000 feet.

VIEW OF THE ALPS.

91. The *Alps* extend from about the meridian of 15° east longitude in a semicircular sweep of about 700 miles to the head of the Gulf of Genoa. Their breadth varies from 100 to 130 miles. They are highest in the western part, where the crest of the range has an average elevation of between 8,000 and 9,000

feet. Mont Blanc, their loftiest summit, and the highest mountain in Europe, has an elevation of 15,760 feet. Numerous summits exceed 10,000 feet in height, and rise above the limit of perpetual snow, the line of which is here between 8,000 and 9,000 feet above the level of the ocean. The most frequented pass, that of Mont Cenis (between Savoy and Piedmont), is 6,770 feet above the sea. It is much more steep and difficult on the Italian side than on that of Savoy. It consists of a plain, 6 miles long by 4 miles wide, encircled on all sides by the different eminences and ridges that form the summit of this part of the chain. The surrounding heights are from 2,500 to 4,500 feet above the plain. The pass of the Great St. Bernard

DOGS OF ST. BERNARD.

is 8,170 feet high. It is celebrated for the passage of the French army over it in the year 1800, but more so for its hospice and sagacious dogs, employed in the rescue of travelers, benighted or endangered by the snow-storms. The pass of Mont Cervin, farther to the eastward, is 11,100 feet, and is the highest pass in Europe, but is not practicable for carriages. The great road of the Simplon, constructed by Napoleon, attains an elevation of 6,585 feet.

92. The *Apennines* commence near the head of the Gulf of Genoa, and extend in a southeasterly direction through Italy. The average height of the crest of the Apennines varies from 3,000 to 5,000 feet, but in the central portion of the chain several summits are between 7,000 and 10,000 feet high. A volcanic region extends along the west side of the Apennines between the 40th and 43d parallels, and at its southern extremity is Mount Vesuvius, 3,932 feet in height, and the only active volcano in continental Europe.

Questions.—87. Where do the mountains of Guiana extend? By what name is the most western of these mountains distinguished? The eastern? 88. What is said of the mountains of Brazil? 89. How many important mountain systems does continental Europe embrace, and where are they situated? What ranges does the south mountain region include? Through what countries does the north mountain region extend? What is said of the Ural and Caucasus mountains? 90. Where are the Balkan Mountains situated?

Questions.—What branch on the north? What branches on the south? Height of the Balkan Mountains? 91. Describe the situation of the Alps. Their breadth? Average height in the western part? What is said of Mont Blanc? Of numerous other summits? What is the height of the limit of perpetual snow? What is said of the pass of Mont Cenis? Great St. Bernard? Mont Cervin? The great road of the Simplon? 92. What is said of the Apennines? Height? Volcanic region? Mount Vesuvius?

93. The *Carpathian Mountains* are situated to the north of the river Danube. The higher portions have an elevation of between 5,000 and 6,000 feet. A number of peaks, however, exceed 8,000 feet. The Carpathians are, in general, exceedingly rugged, and the passes through them narrow and difficult. To the west of the 18th meridian are several ranges, encompassing the plains of Bohemia, sometimes known as the *Hercynian Mountains*, having an average elevation of from 2,000 to 3,000 feet. The range on the north of Bohemia is called the *Erz Gebirg*, words signifying " ore mountains."

94. The mountains of the Spanish peninsula consist of the *Pyrenees* and the *Cantabrian Mountains* in the north; the *Castilian Mountains*, the *Mountains of Toledo*, the range of the *Sierra Morena*, which extend in nearly parallel courses through the central part; and the range of the *Sierra Nevada* in the south. The Pyrenees are about 300 miles in length, and have an average elevation of from 7,000 to 9,000 feet. Their highest summit, the Peak of Nethou, is 11,168 feet in elevation. The Cantabrian Mountains have an average elevation of from 4,000 to 6,000 feet, but some of the summits exceed 10,000 feet. The mountains of central Spain are not remarkable for great height, but few rising above 5,000 feet. The highest summits of the Spanish peninsula are those of the Sierra Nevada, the general height of which varies from 6,000 to 9,000 feet. The Peak of Mulhacen, in this range, has an elevation of 11,657 feet.

95. The mountains of Norway and Sweden, sometimes called the *Scandinavian Mountains*, extend along the Atlantic coast, from the Naze (the south point of Norway) to the North Cape. Their average height is from 3,000 to 5,000 feet, but some peaks, as the Snee-hætten (*snow hat*), in the Dovre-field, are above 8,000 feet in height. The North Cape, in the island of Mageroe, which is a detached member of this mountain system, is a high mass of rock rising to 1,161 feet in elevation, and broken into pyramidal cliffs by the force of the waves.

96. The *Ural Mountains* form the eastern boundary of Europe. They extend from about the 51st parallel, a distance of more than 1,200 miles, to the vicinity of the Arctic Ocean, Their mean elevation is about 2,000 feet. The chain of *Mount Caucasus* extends between the Caspian and Black seas through a length of more than 700 miles. The breadth of the whole mountain region exceeds 150 miles. The highest summit, called El-burz (*the mountain*), is 18,493 feet in elevation. The line of perpetual snow in the Caucasus is about 11,000 feet above the level of the sea.

97. ASIATIC MOUNTAINS.—Asia is remarkable for embracing the most extensive mountain system in the world. The ranges extend mostly in an east and west direction, from the eastern extremity of the continent to the shore of the Mediterranean. Among the principal chains are, the Aldan or Stanovoi Mountains, the Altai, the Thian-shan, the Kuen-lun, the Himalaya, the Hindoo-Koosh, the mountains of Armenia, and the chain of Mount Taurus.

98. The *Himalaya Mountains*, which border the plateau of Tibet on the south, contain the loftiest summits on the globe

(68). This range is about 1,500 miles in length, and from 200 to 250 in breadth, and has a mean elevation of from 15,000 to 18,000 feet. All the higher parts of the mountains are covered with perpetual snow. The mean height of the snow-line is about 15,000 feet on the southern and 18,000 feet on the northern side of the principal range. Some of the passes over the Himalaya are at the remarkable elevation of 18,000 feet above the sea, and several exceed 15,000 feet.

99. The *Altai Mountains*, which extend in an east and west direction, between the 50th and 55th parallels, border the great central table-lands of interior Asia upon the north. The *Thian-shan* and the *Kuen-lun* ranges are intermediate to the Altai and Himalaya ranges. To the east of the Great Desert of Shamo or Gobi are the mountains of *In-shan* and *Khin-ghan*. *Pe-ling* and *Nan-ling* (or northern and southern mountains) extend in an east and west direction through China, separated by the basin of the Yang-tse-kiang River.

100. The mountains of the *Hindoo-Koosh* (the summits of which are from 18,000 to 20,000 feet above the level of the sea) form a group in which several ranges unite : this region joins the elevated plateaus of central Asia with those in the western part of the continent. The *Paropamisan* extend west from the Hindoo-Koosh along the northern borders of the plateau of Iran. The *Elburz Mountains* are a continuation of the same range, and extend south of the Caspian Sea, rising in *Mount Demavend* to a height of 21,500 feet. The *Soleimaun* (or *Suleymaun*) *Mountains* are on the east, and the *Zagros Mountains* on the west of the plateau of Iran.

101. The *Mountains of Armenia* are very irregularly disposed, and are situated between the head waters of the Tigris and the Caspian Sea. Among the highest of these mountains is *Mount Ararat*, which rises to 17,260 feet above the level of

MOUNT ARARAT.

the sea, and is covered with perpetual snow. The chain of *Mount Taurus* extends through Asia Minor, and has an average elevation of from 4,000 to 5,000 feet; its highest summit, *Mount Argœus*, is 12,869 feet above the sea. The *Mountains of Lebanon* extend along the coast of Syria in a north and south direction. *Mount Hermon*, their loftiest summit, is 10,000 feet above the sea-level, and borders on the region of perpetual

Questions.—93. The Carpathian Mountains? Height? General character of the Carpathian Mountains? Ranges to the west of the 18th meridian? 94. What are the principal mountain ranges of the Spanish peninsula? The Pyrenees? The Peak of Nethou? The Cantabrian Mountains? Mountains of central Spain? Highest summits? The Peak of Mulhacen? 95. The mountains of Norway and Sweden? Average height? The North Cape? 96. The Ural Mountains? Extent? Mean elevation? Mount Caucasus? El-burz? Line of perpetual snow? 97. For what is Asia remarkable?

Questions.—General direction of the ranges? Principal chains? 98. Himalaya Mountains? Their length and breadth? Mean elevation? Snow-line? Passes? 99. The Altai Mountains? Thian-shan and Kuen-lun ranges? What mountains to the east of the Great Desert of Shamo? The Pe-ling and Nan-ling? 100. The Hindoo-Koosh? What regions do they unite? The Paropamisan? The Elburz Mountains? The Soleimaun Mountains? The Zagros? 101. The mountains of Armenia? Mount Ararat? Mount Taurus? Argœus? The mountains of Lebanon? Mount Hermon? The Sinai Mountains?

snow. The highest peak of the *Sinai Mountains*, which stand at the head of the Red Sea, is 9,300 feet above the sea.

102. Mountain ranges also extend through Arabia, Hindoostan, the Indo-Chinese peninsula, and the peninsulas of Corea and Kamtchatka.

103. AFRICAN MOUNTAINS.—A range of mountains extends along the northern shores of Africa called *Mount Atlas*. It has a mean elevation of from 7,000 to 9,000 feet; but *Mount Miltsin*, to the southeastward of the city of Morocco, is found to be 11,400 feet in altitude. To the eastward of the 4th meridian of west longitude is a series of ranges nowhere more than from 3,000 to 4,000 feet high.

104. The *Mountains of Abyssinia* constitute another mountain system of Africa. They rest on an extensive plateau of from 6,000 to 8,000 feet in elevation. The highest summits are upward of 15,000 feet above the sea, and are covered with perpetual snow. The valley of the Nile, through nearly the whole length of its course, is bordered by high rocks or hills of an elevation which rarely exceeds from five to six hundred feet. Near the Red Sea is a succession of mountain groups, some of which reach from 6,000 to 9,000 feet in altitude.

105. A third mountainous system of Africa extends along the western coast, between the parallels of 14° of north and 18° of south latitude. Those which lie in an east and west direction, north of the Gulf of Guinea, are known by the name of the *Kong Mountains ;* their general height is from 3,000 to

in the marshy lowlands, combined with other causes, has prevented anything like a full survey of the coast regions of Africa, while the vast interior is almost entirely unknown. It is probable that extensive elevations may yet be discovered in the central part.

106. A fourth series of mountain chains extends along the eastern coast of Africa, though lying generally at a considerable distance inland. But very little is known respecting this range. It is supposed to form the border of a great interior table-land. *Mount Kilimandjaro*, in latitude 4° south, was discovered by a missionary in 1849 ; its summit is covered with perpetual snow, whence its elevation is assumed to be about 20,000 feet. *Mount Kenia*, a volcano, situated a little south of the equator, is apparently somewhat higher than Kilimandjaro.

107. In South Africa is a mountain chain which runs in a general direction of east and west, called the *Nieuveldt Mountains*. The highest portions are above 10,000 feet in height. The *Table Mountain*, in the neighborhood of the Cape of Good Hope, is a flat-topped summit, 3,582 feet above the level of the sea.

NOTE.—*Before proceeding to the following lesson, the student should turn to pages 87–95, and attend to the questions* RELATING TO PLATEAUS AND PLAINS, *on the several maps of the grand divisions. These questions may be assigned for two lessons.*

BARREN PLAIN, EAST OF ROCKY MOUNTAINS.

4,000 feet. The *Peak of Cameroons*, near the coast of the Bight of Biafra, is a detached mountain mass, 13,760 feet in height. In the adjoining island of Fernando Po, *Clarence Peak* rises to 10,655 feet. The extreme unhealthiness of the climate

LESSON IX.

UPLAND PLAINS, OR TABLE-LANDS.

108. UPLAND plains, sometimes called table-lands, or plateaus, are extensive tracts whose general level is considerably elevated above that of the sea. They are commonly skirted by mountain ranges, which in some cases descend abruptly into the surrounding plains. Some of these elevated plains support upon their surface large mountain ranges, which, although of inconsiderable height above the plateau, are yet of great elevation above the level of the sea. Table-lands vary in height from 2,000 to 15,000 feet.

109. The most extensive table-lands of North America are those situated in Mexico. They embrace the plateau of Anahuac, and extend from about the 42d parallel southward to the peninsula of Tehuantepec. This vast highland consists of a series of plains of different elevations : the highest part of the plain of Toluca, upon which the city of Mexico stands, is 9,000 feet above the sea-level : farther to the southeast are the plateaus of Guatemala and Honduras, which exceed 6,000 feet.

Questions.—102. Other mountain ranges ? 108. What mountains in the north of Africa ? Mean elevation ? Mount Miltsin ? 104. On what do the mountains of Abyssinia rest ? Height of the highest summits ? The valley of the Nile ? Mountains near the Red Sea ? 105. What is a third mountain system ? The Kong Mountains ? The Peak of Cameroons ? Clarence Peak ? 106. A fourth series of mountain chains ? What is supposed respecting it ? Mount Kilimandjaro ? Mount Kenia ?

Questions.—107. What mountain chain in South Africa ? Greatest elevation ? The Table Mountain ? 108. What are upland plains, table-lands, or plateaus ? How commonly skirted ? Mountain ranges supported by them ? Height of table-lands ? 109. Table-lands of North America ? What large plateaus do they embrace, and where do they extend ? Of what does this vast highland consist ? Elevation of the plain of Toluca ? Of the plateaus of Guatemala and Honduras ?

110. The plateau of Chihuahua (to the northward of the 24th parallel) varies from 4,000 to 6,000 feet in height, the southern portion being the more elevated. It is generally level, and a great part of it desert. This tract contains many dry salt-lakes, and most of the rivers which cross it terminate on the table-land without finding any outlet to the coast. The plain of Ana-huac is more generally fertile, though arid in many parts.

111, Along the base of the Rocky Mountains a succession of barren plains extends eastward, for a distance of 400 miles, toward the valley of the Mississippi. Through these regions the Red, Arkansas, and other rivers flow in channels consid-erably below the general surface. The country between the Rocky Mountains and the Pacific consists, for the most part, of high plateaus and mountain-terraces, including the Great Basin or plateau of Utah, with an elevation of from 4,000 to 5,000 feet.

112. The northern regions of North America are character-ized by highlands, though of no very considerable elevations (79). The severity of the climate has prevented anything like a full exploration, and they will long remain, as they are at present, inhabited only by a hardy race of savages.

113. South America is remarkable for containing some of the most elevated plains in the world. That which surrounds Lake Titicaca is 12,900 feet above the level of the sea, and is sur-rounded by the loftiest summits of the Andes. The plain of Antisana, under the equator, surrounds the cone of a mount-ain by the same name, which is covered with eternal snow, and seems like an islet in its midst. This plain is 13,451 feet high, and contains the hamlet of Antisana, which lies at the foot of the cone, being one of the highest spots inhabited by man.

114. The most extensive table-land is that of Spain. This peninsula consists chiefly of an elevated tract which reaches on its north side to a height of 3,000 feet, and on its south to about 2,000 feet. Madrid, the capital, has an elevation of 2,170 feet. The plateau of Bavaria, in central Europe, reaches a height of 2,000 feet; and the south-west portion of the Scandinavian peninsula constitutes a pla-teau of moderate elevation. There are several small plateaus, such as the plateau of the Valdai, the plateau of Finland, etc.

115. Asia contains the most widely extended system of table-lands on the globe. Between the Himalaya Mountains on the south, and the Altai Mountains on the north, extends the vast central plateau, having an elevation of from 3,000 to 12,000 feet above the sea-level. In the south, the plain of Tibet attains the remarkable height of 12,000 feet. Farther north is the great desert of Shamo or Gobi, which has an elevation of from 4,000 to 5,000 feet. Nearly the whole of southwestern Asia is elevated into table-lands, among which are the plateau of Iran, the plateau of Asia Minor, and the plateau of Arabia. These plateaus have an elevation of from 2,000 to 4,000 feet. The plateau of the Deccan forms the southern portion of Hindoostan, and has an elevation of about 2,000 feet.

116. The Sahara (or Great Desert) of Africa forms a pla-teau of moderate elevation, probably not more in general than from 1,000 to 1,500 feet above the sea, though particular por-tions of it reach the height of 2,000 feet. It is generally

SCENE IN THE DESERT OF ARABIA.

believed that a vast table-land stretches from the borders of Soudan to Cape Colony; and so far as explorations have been made within these limits, this belief is fully confirmed.

LESSON X.

LOWLAND PLAINS.

117. LOWLAND plains are distinguished from plateaus by being only slightly elevated above the sea-level. In some cases they are considerably below it, as, for examples, the regions around the Caspian Sea and the Sea of Aral. They embrace the most fertile regions of the globe, and being commonly traversed by navigable rivers, affording facilities for inland commerce and communication, they constitute the principal abodes of man, and the seats of industry and wealth.

118. NORTH AMERICAN PLAINS.—The central portion of North America, extending from the Gulf of Mexico to the Arctic Ocean, forms one immense plain, estimated to possess over two and a quarter millions of square miles, or nearly one third of the area of the entire continent. A rising ground di-vides it into a northern and southern slope—the former being drained by the waters which flow into Hudson Bay and the Arctic Ocean, and the latter by streams tributary to the Gulf of Mexico.

119. This plain is bounded on the east by the Apalachian system, and on the west by the highlands which form the east-ern base of the Rocky Mountains. Its western limit, south of the 55th parallel north, is nearly along the 101st meridian; above that parallel it pursues a northwest direction to the mouth of the Mackenzie River, in latitude 135° west.

120. The only considerable elevations throughout this im-mense extent are the Ozark Hills in the south, and a plateau of moderate height to the north and west of Lake Superior.

Questions.—110. Describe the plateau of Chihuahua. What is said of its lakes and rivers? The plain of Anahuac? 111. Plains east of the Rocky Mountains? Between the Rocky Mountains and the Pacific? 112. Highlands in the northern part of North America? 113. For what is South America remarkable? What is said of the plain which surrounds Lake Titicaca? The plain of Antisana? Hamlet of Antisana? 114. The most extensive table-land of Europe? Its elevation on the north and south? Elevation of Madrid? The plateau of Bavaria? Other plateaus? 115. Plateaus of Asia?

Questions.—Describe the situation of the vast central plateau. Its elevation? The plain of Tibet? The great desert of Shamo? Plateau of southwestern Asia? Plateau of the Dec-can? 116. Sahara? Table-land of South Africa? 117. How are lowland plains distin-guished from plateaus? In what case are they below the sea-level? What do they embrace? 118. What is the situation and extent of the great central plain of North America? How divided? 119. Its boundary on the east and west? Its western limit? 120. Elevations? The eastern parts? The middle parts?

The eastern part of the plain, toward the base of the Alleghany Mountains and the shores of Hudson Bay, are generally wooded, and diversified by hills of trifling elevation. The middle parts, embracing the valleys of the Mississippi and Missouri, and the intervening tracts as far as the upper portions of the Mackenzie, are level and grassy regions, called *prairies*.

121. Prairies may be arranged into three kinds:

(1.) The *Bushy Prairies*.—These usually contain springs, and are covered with grass, shrubs, grapevines, and varieties of flowers.

SCENE ON AN AMERICAN PRAIRIE.

(2.) The *Dry* or *Rolling Prairies*, so called from the absence of swamps or pools, and on account of their wavy surface. The vegetation consists principally of grass, weeds, and flowers, which grow with great luxuriance. Over these prairies the American buffaloes roam, in herds of from 40,000 to 50,000.

(3.) The *Moist* or *Wet Prairies*, the smallest division, abound in pools without issue, left by the floods of the rainy seasons. They are covered with a rich vegetation of tall rank grass,

122. Along the Atlantic coast, between the base of the Apalachian Mountains and the sea, stretches a plain, or a comparatively level region, narrow in its northern portion, but increasing to a width of about 250 miles toward its southern limits, as the mountains recede from the coast. Extensive swamps line the coast in several places, and the soil near the sea-shore is frequently sterile; but farther inland the country improves, and contains many fertile tracts. The peninsula of Florida, which belongs to this region, is low and flat, and a large portion of it covered with swamps.

123. SOUTH AMERICAN PLAINS.—A great central plain stretches through the whole length of South America. It is divided into three principal portions—the plain of the Orinoco, the plain of the Amazon, and the plain of the Rio de la Plata,—deriving their names from the three principal rivers by which they are respectively drained. These divisions are distinguished by the names of llanos, selvas, and pampas.

124. The *Llanos*, or *Level Fields*, are those vast plains in Venezuela and New Granada which extend to the north and west from the Orinoco. They have a very level surface, and so gently do they slope toward the sea, that a slight rise in the Orinoco reverses the current of the tributary streams. At the close of the rainy season the llanos are covered with luxuriant grass, and form rich pasture grounds; but during the dry months which succeed, the vegetation is entirely destroyed, and the parched ground opens in deep and wide crevices, giving the whole country the aspect of desolation and sterility.

125. The *Selvas*, or *Forest Plains*, extend over the lower portion of the basin of the Amazon, and within the limits to which the annual inundations of that river and its tributaries extend. A large part of this country is covered with dense forests.

126. Mrs. Somerville thus describes the selvas of South America:

"The soil, enriched for ages by the spoils of the forest, consists of the richest mold. The heat is suffocating in the deep and dark recesses of these primeval woods, where not a breath of air penetrates, and where, after being drenched by the periodical rains, the damp is so excessive that a blue mist rises in the early morning among the huge stems of the trees, and envelops the entangled creepers stretching from bough to bough. A death-like stillness prevails from sunrise to sunset; then the thousands of animals that inhabit these forests join in one loud, discordant roar, not continuous, but in bursts. The beasts seem to be periodically and unanimously roused by some unknown impulse, till the forest rings in universal uproar. Profound silence prevails at midnight, which is broken at the dawn of morning by another general roar of the wild chorus. Nightingales, too, have their fits of silence and song; after a pause they

'——all burst forth in choral minstrelsy,
As if some sudden gale had swept at once
A hundred airy harps.'—COLERIDGE."

127. The *Pampas*, or *Flats*, are immense level plains, variously covered with long, coarse grass, mixed with wild oats, clover, and other herbage. The country between the 32d and

LLANO OF SOUTH AMERICA.

39th parallels consists of swampy tracts, overgrown with canes and tall reeds, and in other districts covered with gigantic thistles, which grow to the height of eight feet, and so thick as

Questions.—121. Into how many kinds may prairies be ranged? Describe the Bushy Prairies,—the Dry, or Rolling Prairies,—the Moist, or Wet Prairies. 122. Describe the plain of the Atlantic coast. The peninsula of Florida. 123. Into how many portions is the great central plain of South America divided? By what names are they distinguished?

Questions.—124. Where are the Llanos situated? What is said of their level surface? How covered during the rainy season? Their aspect during the dry season? 125. Situation of the Selvas? How covered? 126. Mrs. Somerville's description of the selvas of S. America? 127. What are the Pampas? Describe the country between the 32d and 39th parallels.

literally to render the country impassable. During nine months of the year the thistles are here the predominant (and almost the only) feature of the vegetable kingdom, but with the heats of the summer they become burned up, and their tall, leafless stems are leveled to the ground by the powerful blast of the *pampero*, or southwest wind, from the snowy ranges of the Andes, after which the earth is covered for a brief time with herbage. This is destined, with the return of spring, again to give way to the stronger vegetation, which it had succeeded, and for a time supplanted.

128. The plain which extends from the banks of the Negro (latitude 39° south) to the southern extremity of the continent is for the most part barren; in some places it is covered with sand, mixed with stones and gravel. It contains no trees, but a scanty vegetation of shrubs and herbage is found in a few hollows and ravines.*

LESSON XI.

GLACIERS.

129. A GLACIER† is a field or immense mass of ice formed in deep but elevated valleys, or on the sides of mountains. Glaciers occur very extensively among the Alps, and the mountains of Norway, and among the Andes of Patagonia and Terra del Fuego. They are also common among the highlands and upon the precipitous coasts throughout the polar regions. They occur, however, in the greatest numbers and to the greatest extent in the mountains of Switzerland. Along the central part of the Alps, from Mont Blanc to the frontiers of Tyrol, there are reckoned more than 400 glaciers, some of them only 3 miles in length, while others are from 18 to 21 miles long, from 1 to 2¼ miles wide, and from 100 to 600 feet thick. Altogether the glaciers of Switzerland are estimated to cover a surface of upward

VIEW OF A GLACIER.

of 1,000 square miles, and form a sea of ice from the inexhaustible reservoirs of which some of the principal European rivers are supplied.

130. Glaciers are not composed of solid ice, but consist of a mixture of ice, snow, and water. They owe their origin to the accumulation of the snow which falls from the sides of the mountains, and becomes only partially melted during the short summer of these elevated regions.

131. It is a remarkable fact respecting glaciers that they have an onward flow like rivers. The movement is very slow,

not amounting usually to more than a foot in twenty-four hours. Like that of rivers, it is most rapid in the center, and slower at the sides and bottom on account of friction. M. Agassiz obtained the following results, in 1841 and 1842, upon this point:

Annual Motion.
Finster Aar.—Stake nearest the center of the glacier..... 269 feet,
" Stake nearest the side of the glacier....... 160 "
Lauter Aar.—Stake nearest the center .... 245 "
" Stake nearest the side .... 125 "

132. The rate of motion depends upon the seasons; thawing weather, and a wet state of the ice, conducing to its advancement, while cold, whether sudden or prolonged, checks its

MER DE GLACE, SWITZERLAND.

progress. The ice near the shore of the *Mer de Glace* (Sea of Ice), near Chamouni, was found to move as follows:

From June 29 to Sept. 28...............................132 feet.
" Sept. 28 to Dec. 12.............................. 70 "
" Dec. 12 to Feb. 17.............................. 76 "
" Feb. 17 to April 4.............................. 66 "
" April 4 to June 8.............................. 88 "

133. Glaciers, originating in the regions of eternal frost, descend far below the line of perpetual snow. The lowest limit to which perpetual snow extends downward in the Swiss Alps is about 8,500 feet above the level of the sea. The lower glacier of the Aar descends more than 1,500 feet below the snow-line, while others descend 4,000 or 5,000 below the region of perpetual snow and ice, as, for examples, the glaciers of the Upper and Lower Grindelwald.

134. The appearance of glaciers is thus described by Lyell:

"When they descend steep slopes and precipices, or are forced through narrow gorges, the ice is broken up and assumes the most fantastic and picturesque forms, with lofty peaks and pinnacles projecting above the general level. These snow-white masses are often relieved by a dark background of pines, as in the valley of Chamouni; and are not only surrounded with abundance of the wild rhododendron in full flower, but encroach still lower into the region of cultivation, and trespass on fields where the tobacco-plant is flourishing by the side of the peasant's hut."

Questions.—128. Describe the plain which extends to the south of the Negro. 129. What is a glacier? Where do glaciers occur extensively? Where else are they common? Number of the glaciers along the central part of the Alps? Their extent?

* Hughes' Manual of Geography.
† GLACIER is from the Latin *glacies*, ice. The French word *glaciere*, from *glace*, signifies an *ice-house*.

Questions.—Number of square miles covered by the glaciers of Switzerland? 180 Of what do glaciers consist? Their origin? 131. Remarkable fact respecting glaciers? Rate of movement? In what part most rapid? Results obtained by Agassiz? 132. Upon what does the rate of motion depend? State the movement of the Mer de Glace at different seasons. 133. Do glaciers descend below the limit of perpetual snow? What is the lowest limit of perpetual snow on the Alps? To what distance below this line are glaciers known to descend? 134. Appearance of glaciers as described by Lyell?

135. Snow mountains and glaciers, though devoid of vegetation in the upper regions, and presenting a picture of desolation on a scale of magnificence which makes it awful, are yet a striking instance of the truth that nothing was made in vain.

"Herds of chamois are at home amid the frozen heights of the Alps; the Tibetian cow can only bear the climate of the valleys in winter; nor can man pronounce such districts barren, though cheerless in appearance, and never intended for his residence. They minister to his comfort, happiness, and even luxury, as the inexhaustible sources of those streams which in summer, when other waters are evaporated and dried up, roll on through the plains, fountains of fertility and plenty. The Rhone, Rhine, Po, Reuss, Ticino, Aar, Adige, Inn, and Drave, respectively flowing to the German Ocean, Mediterranean, Adriatic, and Black seas, are fed from the snows and glaciers of the Alps."—Rev. Thomas Milner.

LESSON XII.

SNOW MOUNTAINS AND AVALANCHES.

136. The polar regions are covered with fields of perpetual ice and snow. In the temperate and torrid zones everlasting frosts prevail only on the high lands. The elevation at which perpetual snow begins in these zones varies with the latitude. It is 16,000 feet from the level of the sea at the equator; 9,000 feet near latitude 45°; 5,000 feet at latitude 60°; 1,000 feet at latitude 70°; and about latitude 80° the line of perpetual snow comes down to the sea-level.

137. The principal localities of permanent snow in the torrid

MOUNT EGMONT.

and temperate zones are Iceland, Norway, the Alps, and Pyrenees, in Europe; the Caucasus, Himalaya, Kuen-lun, and Altai mountains, in Asia; the range of the Greater Atlas, in Africa;

and the Andes, and higher parts of the Rocky Mountains, in America. The mountains of South Australia, and Mount Egmont, in New Zealand, are snow mountains. The Himalaya derives its name, "the dwelling of snow," from the vast surface occupied by it.

138. The summits of these lofty mountains accumulate enormous masses of snow, which are often precipitated into the surrounding valleys, producing terrible disasters. These descending bodies increase in volume by the dislodgment of other masses, and fall with tremendous velocity and violence, uprooting trees, overwhelming houses and villages, and stopping the flow of streams and rivers.

139. Such falls are very common in Switzerland, where they are called avalanches,* or lavanges. In Norway they are called the snee-fond. Four kinds are noticed—drift, sliding, creeping, and ice avalanches.

140. (1.) A drift avalanche is the fall of the drifts and other accumulations of snow from the upper regions. During calm weather snow collects in enormous volumes on the declivities, where it remains until the wind forces it from its resting-place, and it rushes down into the lower regions. In its progress it forces off other masses, which augment its size, and becoming thus enlarged, it descends with constantly accelerated energy, occasioning as much damage by the whirlwind rush of the air as by the direct attack of the snow.

141. (2.) A sliding avalanche is a descent of snow masses which have become loosened by the heat of the earth. Avalanches of this class occur in spring, and commonly originate in the middle region of mountains. (3.) Creeping avalanches originate in a similar manner, but on less steep declivities. (4.) Ice avalanches are parts of a glacier, detached by the summer heat, or broken off by their own weight, on the extremity projecting over the edge of a precipice.

142. The following interesting description of avalanches is by the Rev. Thomas Milner:

"On descending the Sheideck into the valley of Grindelwald, canton of Berne, the extraordinary effects of an ice avalanche that fell some years ago are observable. The ground is entirely cleared; the trees have been swept away like reeds; an area of at least a mile and a half square is strewn with stones and stumps; a fine forest growing on each side of the area, which was untouched by the falling mass.

143. "A similar avalanche descended near the village of Randa, in one of the valleys of the Valais canton, in 1819. It covered with ice, rubbish, and fragments of rock an area of 2,400 feet in length, by 1,000 feet wide, to the depth of 150 feet. It fell on an uninhabited spot, but the adjoining village was destroyed by the tremendous rush of the compressed air consequent upon the descent of such an enormous mass, about 9,000 feet. Beams of houses were carried nearly a mile into the forest, and the massive stone steeple of the church was snapped asunder.

144. "In the year 1749, a creeping avalanche of snow descended in the valley of Tawich, in the canton of the Grisons, and buried the whole village of Bueras, pushing it at the same time from its site. The catastrophe occurred in the night, and so stealthily, that it was unperceived by the inhabitants, who, on awaking in the morning, were surprised at the prolonged darkness. Sixty out of a hundred persons were dug out alive, obtaining a sufficient supply of air through the interstices of the snow to sustain life.

* Avalanche, from the French avaler, to descend.

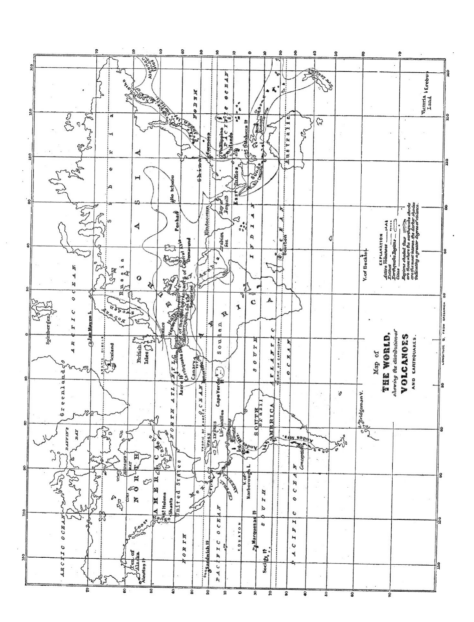

Map of
THE WORLD,
showing the distribution of
VOLCANOES
AND EARTHQUAKES.

145. "In 1838 the secluded hospice of the Grimsel was the scene of a remarkable preservation. The hospice, remote from any human habitation, well known to pilgrims in the Oberland, is only tenanted by a single servant with provisions and dogs, from November to March. In the latter month a great storm occurred, and the snow fell incessantly for four days. While occupied with his art of wood-carving, the solitary was alarmed by a mysterious sound in the evening, like the cry of a human being in distress; but no benighted wayfarer appeared on sallying forth with his dog. The sound recurred again. It was one of those signals which frequently precede a grand catastrophe in the Alps—the noise of a mass disturbed and quivering. Suddenly the impression seized him to retreat into the hospice. He went accordingly into an inner room and began to pray, when the avalanche came thundering down, crushing every apartment but the one which was then sanctified by prayer. Its inmate with his dog succeeded in working his way through the snow, and reached Meyringen in safety, firmly regarding his preservation as an answer to his prayer; and to every pious mind it will verify the sacred declaration, 'He that dwelleth in the secret place of the Most High shall abide under the shadow of the Almighty.'"

LESSON XIII.

VOLCANOES.

146. THE term volcano (derived from Vulcanus, the name the ancient Romans gave to their imaginary god of fire) is applied to those mountains which send forth, from their summits or sides, flame, smoke, ashes, and streams of melted matter called lava: it is also applied to mountains having eruptions of mud only, and which are hence called Mud Volcanoes.

147. Volcanoes are generally of a conical shape, with a hollow at their summit, called the crater, or cup, the sides of which are sometimes entire, like the walls of a circus, but more commonly

VOLCANO OF ORIZABA.

rent. The bottom or floor of craters usually presents a series of ashy cones, with cracks and fissures, through which jets of smoke, steam, and flame issue at the most tranquil intervals.

148. Volcanoes are either continuously active, or intermittent, or extinct. Of the former class is Stromboli, in the Mediterranean—a comparatively lowly mound, 2,175 feet high. It has been uninterruptedly active from the dawn of authentic history, a permanent fiery beacon to the sailors of the adjoining seas, but very rarely violent. Etna, Vesuvius, and Cotopaxi

have varying intervals of rest, in some instances amounting to centuries. Extinct volcanoes are those whose activity has for ages been suspended. A volcano of this class may be found on the isle of Palma, one of the Canaries.

149. An explosion commonly begins by a dense volume of smoke issuing from the crater, mixed with aqueous vapor and gases; then masses of rock and melted matter are thrown out with great violence, after which lava begins to flow, and the whole terminates with a shower of ashes from the crater. The ashes are often the most destructive, as will be seen by the accounts of volcanoes in a succeeding lesson.

150. There are supposed to be about three hundred volcanoes on the earth, about two thirds of which are situated in islands, the remainder being confined to the continents. Their distribution is stated in the following table:

	On Continents.	In Islands.	Total.
Europe	4	20	24
Asia	17	29	46
Africa	2	9	11
America	86	28	114
Oceanica	—	108	108
Total	109	194	303

NOTE.—*Before proceeding with the following lesson, the student should attend to the map questions on page 19.*

LESSON XIV.

VOLCANIC REGIONS.*

151. VOLCANIC REGIONS OF THE ANDES.—The Andes of South America embrace three extensive volcanic regions. The first is known as the Chilean range, the second as the Peruvian, and the third as the volcanic region of Quito. These regions are separated by extensive tracts, in which no volcanic action has been known to occur since the discovery of America.

152. The volcanic range of Chile embraces the most southern line of active vents. It extends from latitude 43° 28' south, or from a point on a range with the island of Chiloe to Coquimbo, in latitude 30° south—a distance of nearly fourteen degrees of latitude. This region is noted for containing the Peak of Aconcagua (22,301 feet), till recently the highest known volcano in the world. To the north is a space of more than eight degrees of latitude, in which no recent volcanic eruptions have been observed.

153. The Peru-Bolivian line of volcanoes, the next in order, extends a distance of about six degrees, from latitude 21° south to latitude 15° south. The volcano of Sahama, of late said to be the highest known volcano—having an elevation of 22,250 feet, is situated in this district. Between the Peruvian volcanoes and those of Quito another space intervenes, of no less

than fourteen degrees of latitude, said to be free from volcanic action so far as yet known.

154. The volcanic region of Quito begins about two degrees south of the equator, and continues about five degrees north of the line. Its most elevated volcanoes are Antisana and Cyambe,—the former having an altitude of 19,370 feet, and the latter 19,534 feet. To the north of this region there occurs another undisturbed interval of more than six degrees, after which we arrive at the volcanoes of Guatemala, in Central America, north of the Isthmus of Panama.

155. The volcanoes of South America tower in many instances to vast elevations above the level of the sea, several examples of which have been previously mentioned. Those of Peru rise from a lofty platform to heights varying from 17,000 to 20,000 feet. Of those which have recently been in a state of activity, the most lofty is Cotopaxi: its eruptions have been more frequent and destructive than those of any other mountain. It is a perfect cone, usually covered with an enormous bed of snow, which has, however, been sometimes melted suddenly by an eruption; as in January, 1803, for example, when the snows were dissolved in one night.

156. Frequent deluges are caused in the Andes by the sudden melting of great masses of snow, and by the rending open, during earthquakes, of subterranean cavities filled with water. In these inundations fine volcanic sand, loose stones, and other materials, which the water meets with in its descent, are swept away, and a vast quantity of mud, called "moya," is thus formed and carried down into the lower regions. In

VOLCANO OF JORULLO.

1797, valleys a thousand feet wide, surrounding Tunguragua, in Quito, were filled with mud from this source to a depth of six hundred feet.

157. VOLCANOES OF NORTH AMERICA.—Proceeding north from the Isthmus of Panama, we find extensive volcanoes

* The description of the volcanic regions contained in this lesson is derived chiefly from "Lyell's Principles of Geology."

Questions.—149. How does an explosion commonly begin? What substances follow? What finally succeeds? Which of the ejected substances is often the most destructive? What number belongs to Europe? To Asia? To Africa? To America? To Oceanica? 150. Number of volcanoes? How many of these are situated in islands? On continents? What number belongs to Europe? To Asia? To Africa? To America? To Oceanica? 151. Number of volcanic regions in the Andes? By what names known?

Questions.—How are these regions separated? 152. Volcanic range of Chile? Between what parallels does it extend? For what remarkable? 158. Situation and extent of the Peru-Bolivian line of volcanoes? Volcano of Sahama? What non-volcanic region to the north? 154. Situation and extent of the volcanic region of Quito? Most elevated summits? What region extends farther north? 155. What is said of the vast elevations of the South American volcanoes? Of those of Peru? What is said of Cotopaxi? 156. Frequent deluges, how caused? Moya? Illustration? 157. Volcanoes north of the Isthmus of Panama? Coseguina? Amount of ashes thrown out?

scattered through Central America and Mexico. Coseguina, in the former country, was in eruption in 1835, and some of its ashes fell at Truxillo, on the shores of the Gulf of Mexico. The amount of ashes thrown out by this eruption was so great, that twenty-four miles to the southward they covered the ground to the depth of three yards and a half, destroying the woods and dwellings. Thousands of cattle perished, and the streams were strewn with dead fish.

158. Of the volcanoes of Mexico are Tuxtla, Orizaba, Popocatepetl, Jorullo, and Colima, situated on a line extending from west to east, near the parallel of 19° north latitude. If this line be prolonged in a westerly direction, it cuts the volcanic group of islands called the Isles of Revillagigedo. There are said to be three, or, according to some, five volcanoes in the peninsula of California; one in the northern part of the State of California (Mount Shasta); one near the mouth of the Columbia River (Mount St. Helens); also several in the southern part of Russian America.

159. Of the West Indian Islands, the range known as the Lesser Antilles is extensively volcanic. It comprises two parallel series: the one to the west, which are all volcanic, and which rise to the height of several thousand feet; the other to the east, for the most part composed of calcareous rocks, and very low. In the former, or volcanic series, are Grenada, St. Vincent, St. Lucia, Martinique, Dominica, Guadaloupe, Montserrat, Nevis, and St. Eustace. In the calcareous chain are Tobago, Barbadoes, St. Bartholomew, and St. Martin. The most considerable eruptions of modern times have been those of St. Vincent.

160. The volcanic regions of the continental parts of America are confined to the western coast, where a line of volcanoes may be traced—as we have seen,—with occasional interruptions, from one extremity of the continent to the other. There seems to be no indication of volcanoes in Buenos Ayres, Brazil, and the eastern part of the United States; though violent earthquakes have occurred in the latter country; as that which convulsed the valley of the Mississippi, at New Madrid, in 1812.

161. VOLCANIC REGION FROM THE ALEUTIAN ISLES TO THE MOLUCCAS AND ISLES OF SUNDA.—An extensive line of volcanoes ranges through the Pacific, parallel with, and at no great distance from, the eastern Asiatic coast. It commences in the north with the Aleutian Isles, and extends, first in a westerly direction for nearly 200 geographical miles, and then southward, with few interruptions, throughout a space of between sixty and seventy degrees of latitude to the Moluccas. At this point it sends off a branch to the southeast, while the principal train continues westerly through Sumbawa and Java to Sumatra, and thence in a northwesterly direction to the Bay of Bengal.

162. It is supposed the northern extremity of this extensive volcanic region is not far from Cook's Inlet, northeast of the peninsula of Alaska, where one volcano, in about the sixtieth degree of latitude, is said to be over 12,000 feet high. Alaska contains cones of vast height, which have been seen in eruption, and which are covered for two thirds of their height downward with perpetual snow.

163. From Alaska the line is continued through the Aleutian or Fox Islands to Kamtchatka. On this peninsula are many active volcanoes, which, in some eruptions, have scattered ashes to immense distances. Of these, the largest is Klutschen, latitude 56° north, which rises at once from the sea to the prodigious height of 15,763 feet.

164. The Kurile chain of islands constitutes the prolongation of this range, which is continued through Jesso, Niphon, Loo-Choo, and Formosa to the Philippine Islands and the Moluccas.

165. Java is said to contain thirty-eight considerable volcanoes, some of which are more than 10,000 feet high. They are remarkable for the quantity of sulphur and sulphurous vapors which they discharge. They rarely emit lava, but rivers of mud issue from them in great quantities. There are numerous extinct craters on the island of Java filled with water strongly impregnated with sulphuric acid, and the streams flowing from them will support no living creature.

166. The Indian and Pacific oceans contain a great number of volcanic islands, interspersed with those of coral formation. The former are lofty, and present evidence that they have been undergoing upheaval in modern times; the latter are very low, consisting of reefs of coral, usually with lagoons or lakes in their centers.

167. VOLCANIC REGIONS OF THE MEDITERRANEAN.—Of the volcanic regions in, or bordering on, the Mediterranean Sea, are those of Greece and Italy, with the adjacent islands. In the Grecian Archipelago is the island of Santorin, a grand center of volcanic action, and the Ionian Isles are continually convulsed. Vesuvius and Etna,—the former in southern Italy, and the latter in the island of Sicily,—are among the most remarkable volcanoes in the world. The volcanic region which traverses the northern shore of the Mediterranean is supposed to extend in the east through Asia Minor, and the countries bordering on the Caspian Sea to central Asia, and on the west, through the southern part of Spain and Portugal to the Azores.

LESSON XV.

VESUVIUS, ETNA, ETC.

168. VESUVIUS.—This volcano is situated about six miles from Naples, and rises in a pyramidal form from a large plain to an elevation of 3,948 feet. It has been subject to many powerful convulsions, some of which have produced great devastation in the surrounding districts.

Questions.—158. Volcanoes in Mexico? How situated? Volcanoes in the peninsula of California? What volcano reported to have been found farther north? 159. What volcanic range in the West Indian Islands? Describe the more westerly of the two parallel series. The other. What islands belong to the former series? To the other? 160. To what regions are the volcanoes of the continental part of America confined? Where may they be traced? In which countries are there no traces of volcanoes? 161. Volcanic range through the Pacific? Where does it commence? Give its general course. Its course from the Moluccas? 162. Situation of the northern extremity of this volcanic region? What of the volcanoes of Alaska?

Questions.—163. Course of the volcanic line from Alaska? Volcanoes of Kamtchatka? Klutschen? 164. Through what chain of islands is the range prolonged? Through what other islands is the chain continued? 165. Number of volcanoes in Java? For what are they remarkable? Ejected matter? Extinct craters? 166. Extent of volcanoes in the Indian and Pacific oceans? How interspersed? Character of the volcanic islands of the coral islands? 167. Volcanic regions of the Mediterranean? Santorin? Ionian Isles? Vesuvius and Etna? Through what countries does this volcanic region extend on the east? On the west? 168. Vesuvius, where situated? Its form and elevation? Convulsions?

169. The following description of its crater is by an American[*] who visited it in 1851:

"After passing through thick clouds of steam charged with the suffocating sulphurous acid gas which greatly annoyed our lungs, when the wind cleared our vision we found ourselves on the narrow rim of the great crater. It was so narrow that only two persons could walk upon it abreast. . . . . On this narrow rim we walked on a surface not more than six or eight feet wide, with the terrific crater, 1,000 feet deep, on one side, and the abrupt descent of 1,800 feet on the other side, into the valley of Somma. It was a spectacle truly sublime, awfully grand and appalling. The heat below sent up, in throes and spasms, dense clouds of steam and sulphurous acid gas, which, at short intervals, filled the crater, and all that appeared was a pillar of a cloud, in which we were often involved and half suffocated; we stood with arms locked, for then it was unsafe to move until a whirlwind swept the crater clear, and we could see into its profound abyss.

170. "Nothing could be more perfectly formed than this crater. It was a magnificent hollow cone, whose nether apex opened into the great world of fire below; still the fire we did not see, although we had the most decisive proof of its existence and continued action in the violent ejection of steam and gas, filling every few minutes this vast funnel, whose diameter across from side to side was probably 1,000 feet; but the dense cloud of steam and noxious gas which hovered to leeward over a large portion of the circular orifice, rendered it impossible to walk around it, or even to

CRATER OF VESUVIUS IN 1829.

ascertain whether a continuity of surface, free from cross fractures and chasms, would render it practicable without the most imminent danger."

171. The earliest recorded eruption of Vesuvius is that which happened in the year of our Lord 79, and by which the cities of Herculaneum and Pompeii were destroyed.

172. History informs us of that sad catastrophe, but the sites of those cities had become unknown, and it was not until the last century that they were discovered. Herculaneum, which was discovered in 1711, lies five miles from Naples, immediately adjacent to the eastern shores of the bay, and is still partly covered by the large modern village of Resina. Pompeii, discovered in 1748, is situated farther to the southeast, twelve miles distant from Naples. A great part of it has been cleared from the ashes under which it had so long lain buried, and it exhibits to view the full picture of what a Roman city was,—temples; theaters, baths, private habitations, the shops of the different trades, the implements with which they were carried on, and even the materials upon which these were employed.

173. In the vicinity of Naples there are other evidences

* Professor B. Silliman.

of the volcanic character of this district. Numerous hot baths abound, supplied with steam and water from the vents in the

VESUVIUS.

earth. The *Solfatara* is the name of a nearly extinguished volcano, giving vent continually to aqueous vapor, together with sulphurous and muriatic gases, like those which escape from Vesuvius.

174. ETNA.—This celebrated volcano is situated on the eastern side of the island of Sicily. It rises near the sea in solitary grandeur to the height of nearly eleven thousand feet. Its base is almost circular, and is about eighty-seven English miles in circumference. Its summit is about thirty miles from the town of Catania, a sea-port to the south.

175. Etna has been longest known to history of any volcano. It was known to be in action 480 years B.C. Here, according to ancient mythology, the thunderbolts of Jupiter were forged; here, also, was raised a temple to Vulcan, where the fire never ceased to burn. From the above-mentioned date, during the whole subsequent period, it has been subject to periodical convulsions, with intervals of repose varying greatly in duration. The recorded eruptions of Etna are....B.C. 480—427, interval 53 years.

"	"	....B.C. 427—396,	"	31 "
"	"	....B.C. 396—140,	"	256 "
"	"	between B.C. 140—122, four eruptions.		
"	""	"	B.C. 122— 56, interval 66 years.	
"	"	between B.C. 56 & 38, three eruptions.		
"	"	B.C. 38—A.D. 40, interval 78 y's.		
"	"	A.D. 40— 251,	"	211 "
"	"	A.D.251— 812,	"	561 "
"	"	A.D.812—1169,	"	857 "

From this period to the present the agitations of Etna have increased in frequency and power:

Twelfth and thirteenth centuries.....................	3 eruptions
Fourteenth.....................................	2 "
Fifteenth......................................	4 "
Sixteenth......................................	3 "
Seventeenth....................................	8 "
Eighteenth.....................................	14 "
Nineteenth to 1832..............................	6 "

176. The most remarkable eruption of Etna in modern times occurred in 1669. It was preceded by an earthquake, which

leveled to the ground all the houses in Nicolosi, a town situated twenty miles distant. A fissure, six feet broad, and of unknown depth, opened with a loud crash, and ran to within a mile of the summit. A stream of lava flowed down the side of the mountain, and, after destroying fourteen towns and villages, overwhelmed a part of the city of Catania. This mass, at its entrance into the sea, was six hundred yards broad and forty feet deep. The solid contents of this immense stream of lava is estimated at 93,838,590 cubic feet.

MOUNT ETNA.

177. VOLCANOES OF ICELAND.—Iceland, in the North Atlantic, seems to form a volcanic region by itself. From the beginning of the twelfth century there is clear evidence that, during the whole period, there has never been an interval of more than forty, and very rarely one of twenty years, without either an eruption or a great earthquake. So intense is the energy of the volcanic action in this region, that some eruptions of Hecla have lasted six years without ceasing. Earthquakes have often shaken the whole island at once, causing great changes in the interior, such as the sinking down of hills, the rending of mountains, the desertion by rivers of their channels, and the appearance of new lakes. New islands have often been thrown up near the coast, some of which still exist; while others have disappeared, either by subsidence or the action of the waves.*

178. The greatest eruption on record proceeded from Skaptar Jokul, in Iceland, in 1783. The lava flowed in two nearly opposite streams, fifty miles in one direction and forty in the other. The breadth which one branch attained in the low countries was from twelve to fifteen miles, that of the other about seven. The ordinary height of both currents was one hundred feet, but in narrow defiles it sometimes amounted to six hundred. It has been calculated that the mass of lava thrown out during this eruption surpassed in magnitude the bulk of Mount Blanc.

179. The eruption of Skaptar Jokul did not entirely cease till the end of two years; and when the tract were visited eleven years afterward, columns of smoke were found still rising from parts of the lava. The destruction of life and property was immense. No less than twenty villages were destroyed, besides those inundated by water. More than nine thousand human beings—nearly one fifth the entire population—perished, together with an immense number of cattle, partly by the depredations of the lava, partly by the noxious vapors which impregnated the air, and, in part, by the famine caused by showers of ashes throughout the island, and the desertion of the coasts by the fish.

Questions.—177. Iceland? Frequency of eruptions? Energy of volcanic action? Ravages of earthquakes? New islands? 178. Eruption of Skaptar Jokul? Give particulars of the flow of the lava. Breadth and depth of the lava currents? 179. Duration of the eruption? Destruction of life and property? 180. The term Geyser, how derived? To what applied? Irregular action of geysers. 181. Give particulars of the Great Geyser. Describe the eruptions.

* Sir Charles Lyell.

180. GEYSERS.—The term Geyser is from the Icelandic geysa, which signifies to rage, or burst forth impetuously. It is applied to the hot springs which occur in a remarkable group of fifty or more which occur in Iceland, about thirty-six miles from Hecla. Geysers are not constantly active; few of them play longer than five or six minutes at a time, although sometimes half an hour. The grand eruptions are often after intervals of a day or two.

181. The Great Geyser has a basin at its summit sixty feet in diameter, and six or seven deep. At the bottom of the basin there is a well or funnel, ten feet wide at the mouth, but gradually narrowing to seven or eight, with a perpendicular descent of seventy feet. The eruptions are preceded by subterraneous noises like the distant firing of cannon, and shakings of the earth. The sound then increases and becomes more violent, till at length a column of water is thrown up to the height of one or two hundred feet. After the water ceases to play, a column of steam, rushing up with amazing force and thundering sound, terminates the eruption.*

LESSON XVI

EARTHQUAKES.

182. EARTHQUAKES appear to be due to the same cause which occasions a volcanic eruption, namely, the energy of elastic vapors struggling to find a vent from beneath the surface of the earth. They are most common in volcanic districts; but those at a greater distance from volcanoes are more violent, as if the latter afforded passage for the eruptive energy, which, at other points, cracked and upheaved the surface in effecting its disengagement.

183. The best proof that earthquakes and volcanoes have a similar origin is the constancy with which they attend each other. A few examples are selected in illustration.

184. The same night Lima was destroyed by an earthquake, four new volcanic vents were found in the Andes. Soon after the earthquake at Lisbon, in 1755, there happened some of the most violent eruptions that ever afflicted the world. Thirty days after the destruction of the city of Caracas, the volcano of St. Vincent became active, and at the moment it broke forth the earth shaken over an extent of nearly 20,000 square miles.

185. Earthquakes differ greatly in intensity. The agitation is sometimes so weak as to be scarcely sensible; at other times it is so violent as to overturn cities, prostrate trees, turn the course of rivers, and change the entire aspect of a country.

186. The movements of the surface during an earthquake are various—vertical, horizontal, and undulatory, or whirling. Low rumbling noises resembling distant thunder, or sharp sounds resembling the clanking of chains and discharges of artillery, are commonly heard in great convulsions.

187. The violence of an earthquake seldom lasts more than a minute; but successive shocks are sometimes felt at very slow

Questions.—182. To what do earthquakes appear to be due? Where most common? Where most violent, and why? 183. Best proof that volcanoes and earthquakes have a similar origin? 184. What examples are given in illustration? 185. Intensity of earthquakes? 186. Movements of the surface during an earthquake? Noises commonly heard in great convulsions? 187. Violence of an earthquake? What is said of the most destructive earthquakes?

* Sir Charles Lyell.

intervals. The most destructive earthquakes are the shortest in duration, amounting to little more than the paroxysm of a few moments.

188. The great earthquake of Lisbon, November 1, 1775, was over in about six minutes; the three shocks which reduced the city of Caracas to ruins, March 26, 1812, transpired in the space of fifty seconds; and the principal convulsion which leveled the city of Conception with the ground, February 20, 1835, lasted but six seconds.

189. In some countries, earthquakes of greater or less violence occur almost daily; and in order to guard against their destructive effects, the inhabitants seldom build their houses more than one story in height [see view of Caracas]. At Lima,

CARACAS.

on the Peruvian coast, an average of forty-five shocks may be expected in the year. At Coquimbo, on the coast of Chile, there were noticed during one year not less than sixty-one convulsions, not including the slighter ones, which were even more numerous.

190. Among the remarkable phenomena attending earth-quakes may be reckoned the permanent elevation and depres-sion of large areas of land, the opening of extensive fissures, great oceanic waves, etc.

191. The earthquake of Chile, in 1822, agitated the coast for a distance of 1,000 miles. The rise upon the coast was from two to four feet: at the distance of a mile inland it was sup-posed to be nearly twice as many. It has been conjectured that the area over which this permanent alteration of level extended may have been equal to 100,000 square miles.

192. Numerous instances of the depression of land have occurred. During the great earthquake at Lisbon, in 1755, the new quay subsided and its place was occupied by water 600 feet deep. In 1819, a region of 2,000 square miles, near the mouth of the Indus, embracing the Fort of Sindree, was submerged, the Ullah Bund, a mound near by, rising as a compensating elevation.

193. Clefts or fissures are frequently formed by earthquakes, in which houses, trees, animals, and men have been engulfed in an instant; the earth sometimes closing up, and no vestige of them remaining on the surface. The great earthquake of Calabria, in 1783, furnished numerous instances of such chasms.

194. The neighboring waters of the ocean, during an earthquake, are strongly agitated. During the earthquake at Lisbon a great wave swept over the coast of Spain, and is said to have been sixty feet high at Cadiz. At Tangier, in Africa, it rose and fell eighteen times on the coast: at Funchal, in Madeira, it rose fifteen feet above high-water mark.

195. Earthquakes have caused an immense de-struction of life and property, and in some parts of the world, as in South America, the inhabitants are in constant apprehension of danger.

196. During the earthquake which visited Peru in 1746, 3,800 of the in-habitants perished. In 1797 Peru was visited by another earthquake, on which occasion 16,000 persons perished. The earthquake at Caracas, in 1812, destroyed 10,000 inhabitants. The great earthquake of Lisbon, in 1755, destroyed 60,000 persons in the course of about six minutes. The number of persons lost during the earthquake in the two Calabrias and Sicily, in 1783, is estimated at 40,000; and about 20,000 more died by epi-demics which resulted from it.

Questions.—188. Examples? 189. Frequency of earthquakes in some countries? Average annual number at Lima? Number of convulsions noticed in one year at Co-quimbo? 190. Phenomena attending earthquakes? 191. Earthquake of Chile in 1822? Give particulars. 192. Give the examples of permanent land depression.

Questions.—193. What is said of clefts or fissures? 194. Example of oceanic movements caused by earthquakes? 195. Fatal effects of earthquakes? 196. Number of persons who perished during the earthquake of Peru, in 1746?—1797?—the earthquake at Caracas, in 1812?—the earthquake at Lisbon, in 1755?—the earthquake of the two Calabrias, in 1788?

PART II.

THE WATERS.

LESSON I.

CHEMICAL COMPOSITION OF WATER.

ATER, which is necessary to the support of animal and vegetable life, is very widely diffused and copiously supplied. It is found in three forms: vaporous, in the atmosphere, solid in ice and snow, and liquid in rivers and seas. It belongs to this part of physical geography to treat of it in the last condition.

198. Water is composed of two gases, oxygen and hydrogen, in the proportion of eight parts of the former to one of the latter. It is one of the most marvelous facts in the natural world, that, though hydrogen is highly inflammable, and oxygen is a supporter of combustion, both, combined in water, form an element destructive to fire. By processes well known to the chemist, water may be readily resolved into its constituent elements.

199. Pure water is destitute of color, taste, and smell. It seldom occurs, however, in this state, but contains various ingredients, derived either from the atmosphere or from the earth. Rain-water is the purest that can be obtained except by distillation. Spring and well water contain many earthy substances in solution. The brackish taste of wells in countries abounding in limestone is owing to the presence of that substance. River-water has its character determined by the soil and vegetation of the country through which it flows.

200. The waters of the globe are divided into the fresh and salt. The fresh waters include those of all streams and rivers, nearly all the springs, and the greater number of lakes and marshes. They are so called because they contain no amount of saline matter unfitting them for use. It is supposed that the lakes of North America contain more than half the amount of fresh water on the face of the globe.

201. Salt water is that which fills the vast basins of the ocean, besides numerous lakes and springs: it forms by far the largest portion of the liquid element. The proportion of saline matter which the ocean contains is about 3½ per cent. The principal salts contained in sea-water are common salt (chloride of sodium), Glauber's salt (sulphate of soda), Epsom salt (sulphate of magnesia), chloride of magnesium, sulphate and carbonate of lime,—common salt being the most considerable in amount. Supposing the sea to have a mean depth of 1,000 feet, it has been calculated that the amount of common salt it would contain would be equal in extent to five times the mass of the Alps, or one third less than that of the Himalaya Mountains.

202. Oceanic waters vary in the quantity of saline matter they contain in different places. From observations made, it is found that the degree of saltness diminishes toward the poles, and also near the shores. This is owing to the melting of snow and ice, and to the volumes of fresh water poured in by the rivers. It is also ascertained that the waters of the southern hemisphere contain more salt than those of the northern, while the Atlantic is in excess of the Pacific.

203. Salt water has an extensive distribution in lakes and springs, and these are remarkable for the great proportion of saline matter they contain. The western part of Asia and the southern part of Russia constitute the great salt-water lake region of the globe. The Caspian Sea, lakes Aral, Urumiah, Elton, and the Dead Sea occur in this district. Some of these waters are so salt as to irritate the skin. Fish can not

DEAD SEA.

live in them, and if a bird dips in their surface, its wings, on drying, are incrusted with salt. The waters of the Dead Sea contain about 25 per cent. of saline matter. Lake Elton, in the steppe east of the Volga, is more strongly impregnated with saline ingredients than any other known example, containing 29 per cent.

204. Water is one of the most widely diffused bodies of nature. We have seen (12) that the surface of the earth is estimated to contain 196,500,000 square miles, and that the dry land occupies only about 51,000,000—leaving 145,500,000 square miles to be occupied by the fluid element.

205. The benevolence of the Deity is manifest in the wide diffusion of this element over the globe. As a nutritive or alimentary substance, it is indispensable to both the animal and the vegetable world. It serves invaluable purposes in the arts and manufactures; in its application as a motive power; and, when occurring in large bodies, in the form of rivers, lakes, and seas, as a medium for the more rapid or more commodious transport of goods or persons from one locality to another. To the vast reservoir of the ocean are we indebted for the clouds which carry the moisture from the sea

Questions.—197. What is said of water? In what three forms is it found? 198. Of what is water composed? What marvelous fact is stated respecting water? 199. What is the character of pure water? Does it commonly occur in this state? What kind of water is the purest? What do spring and well water contain? To what is the brackish taste of wells in limestone countries owing? By what is the character of river-water determined? 200. How are the waters of the globe divided? What do the fresh waters include? Why are they so called? What is said of the lakes of North America?

Questions.—201. What is said of salt water? What proportion of saline ingredients does it contain? What are the principal salts in sea-water? Which is the most considerable in amount? Illustrate. 202. How do oceanic waters vary? Where is the saltness of the ocean found to diminish? Cause? Between what other regions is the saltness unequal? 203. What is said of salt lakes and springs? Where is the great salt-water lake region? Examples. 204. Diffusion of water? Extent of the liquid element? 205. Mention some of the uses of water.

THE WATERS.

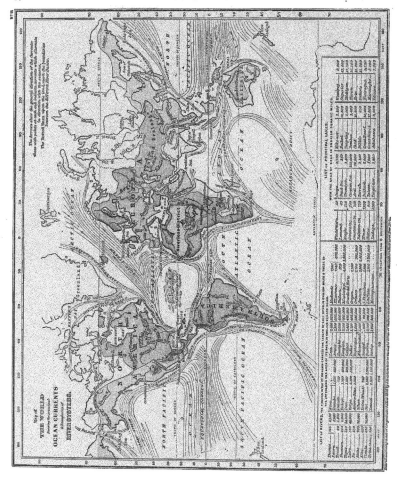

QUESTIONS ON THE MAP.

OCEANS, SEAS, ETC.

What two great oceans are separated from each other merely by a succession of islands? What two are connected solely by a narrow strait? With what part of the Arctic Ocean does nearly half the breadth of the North Atlantic unite?

What ocean is most thickly studded with islands? What is the vast island-region of the Pacific called? *Ans. Oceanica.* Does it extend nearest to America or Asia? What part of the Pacific presents the broadest expanse of ocean uninterrupted with islands to be found on the globe? Near what part of the same line the Antarctic Ocean been explored farthest south? [*See lower right-hand corner of the map.*]

Which of the oceans is bordered by the greatest number of seas separated from it chiefly by islands? Name some of these seas. What ocean has the most seas, gulfs, and bays, communicating with it, which extend far into the continents, or are nearly land-locked? Name some of these seas, etc.

SYSTEMS OF DRAINAGE.

Does the Atlantic Ocean, or the Pacific and Indian oceans taken together, receive the drainage of the greatest area on the continents? [*The extent of drainage belonging to each ocean, or of the river systems tributary to each, is indicated by the coloring.*]

Note.—The student will recollect that in the questions on the preceding map, attention was called to the fact that the short and abrupt slopes of the continents, in consequence of the situation of the great mountain-chains, are toward the Pacific and Indian oceans; the long slopes, in the opposite directions. Hence the system of drainage is as it now appears.

On what continent has the Pacific basin or region of drainage its great extent? On ?, the Atlantic ?, the ? great rivers belonging to the Pacific basin in Asia. To the ? ? in America. To the ? in ? and Africa.

Does the basin of the ? Ocean extend ? on the continents? [*It will be remembered that, as the neighboring map is drawn on Mercator's projection, the northern and southern regions are much enlarged beyond their true relative size. This will be ? on comparing it with a globe or map of the hemispheres.*] On which has it the greater area? Name the great rivers which it includes on the ? Continent. On the Western.

What two grand divisions together contain an immense region of inland drainage? What great salt lakes or seas receive most of the waters of this region? ?? are the ? tributaries of the Caspian?

OCEAN CURRENTS.

What vast ocean-current, interrupted by the continents, flows westward through a great part of the torrid regions? Into what two branches does it divide on the eastern coast of South America? What great current, fed by its northern branch and by the heated waters of the Gulf of Mexico, sweeps across the North Atlantic? Into what ocean is a part of the waters of the Gulf Stream discharged? What is the course of the other part? What extensive bank or collection of floating sea-weed lies between the Equatorial Current and the southern branch of the Gulf Stream? By what other name is this sea-weed region often called? *Ans. Sargasso Sea.*

What current from the Arctic regions flows east of Greenland? What one from Baffin and Hudson bays unites with it, and flows between the Gulf Stream and the American coast?

What current originating in the southeastern flow of the Gulf Stream runs along the northwest coast of Africa? What one from the Gulf of Guinea flows south to the Antarctic Ocean? What branches from the Equatorial Current of the Indian Ocean unite with this?

What branch from the Equatorial Current of the Pacific, near the coast of Asia, has a similar course to that of the Gulf Stream? By what name is this current designated in the northeastern part of the Pacific?

What great current on the west coast of South America is tributary to the Equatorial Current? By what general name is the northeastward flow of waters from various parts of the Antarctic Ocean termed? What is that branch called which sweeps past Cape Horn?

---

and let it down upon the parched and thirsty earth in refreshing rain; indeed, when we study the uses and properties of water, and notice how universally it is supplied to the inhabitants of the earth in the conditions best adapted to their wants and necessities, we can not but be filled with gratitude to Him who so created and distributed it.

Note.—The questions on the neighboring map should receive attention before proceeding with the following lesson.

LESSON II.

MINERAL SPRINGS.

206. MINERAL waters may be divided into four classes: (1) acidulous, such as contain carbonic acid; (2) chalybeate, or springs holding in solution either the carbonate or the sulphate of iron; (3) sulphurous, or springs containing either sulphureted hydrogen, or sulphuret of lime, etc., and which are distinguished by their repulsive smell; and (4) saline, or springs holding in solution a considerable portion of neutral salts, which render them hard, and impart to them a disagreeable taste, unfitting them for a beverage, or for culinary purposes.

207. *Acidulous* waters present a sparkling appearance which they owe to the presence of carbonic acid gas. This gas is very plentifully disengaged from springs in almost all countries, but more particularly in the vicinity of active or extinct volcanoes. Acidulated waters sparkle when poured from one vessel into another. A remarkable spring containing water of this description occurs in Asia Minor, near Kiz-hisar, which rises very cold, but bubbles up like a boiling caldron. A small river in South America, a tributary of the Magdalena, which rises in a volcanic mountain, has its waters so greatly impregnated with this gas, that the Spaniards call it Vinagré (*vinegar*).

208. *Chalybeate* springs are such as contain oxide of iron; the term is, however, occasionally applied to springs containing other substances. The springs at Tonbridge Wells and Brighton, England, belong to this class.

209. *Sulphurous* springs are so named because they contain sulphur, which usually occurs in the form of sulphureted hydrogen, or of sulphate of lime. Of the springs of this class are those of Harrowgate, and St. Bernard's Well, England. Sulphurous springs are very numerous in volcanic countries.

210. *Saline* springs contain a greater or less proportion of chloride of sodium, or common salt; and are very generally diffused throughout the surface of the earth. So great is the quantity of chloride of sodium in some of these springs, that they yield one fourth of their weight in salt. The springs at Salina and Syracuse, in the State of New York, are noted examples of this class. The principal spring at Salina village affords an inexhaustible supply of water, yielding to every forty gallons about one bushel of pure salt. In 1850 there were manufactured at these springs over 8¼ millions of bushels of salt. Salt springs occur also on the Great and Little Kanawha rivers, in Virginia,

Questions.—206. Mineral waters, how divided? Acidulous?, Chalybeate? Sulphurous? Saline? 207. Appearance of acidulous waters? Carbonic acid gas? Spring near Kiz-hisar? A river in South America? 208. Chalybeate springs? Examples. 209. Sulphurous springs? Examples. Where numerous? 210. Saline springs? Quantity of salt they contain? Springs at Salina and Syracuse? On the Great and Little Kanawha? At Droitwich?

which yield about a bushel of salt from every 60 or 70 gallons of brine. At Droitwich, England, is a celebrated brine spring which produces annually about 700,000 bushels of salt.

211. Besides the springs above described, there are others whose waters are impregnated with various mineral substances which impart to them medicinal properties, as those at Cheltenham, England, at Carlsbad, Germany, and at other places.

212. The mineral springs of the United States are numerous and valuable: they are extensively distributed, but those of New York and Virginia are the most celebrated. Among those in the State of New York are, the chalybeate springs of Saratoga, the sulphur springs of Sharon, Avon, and Clifton, and the petrifying springs of Madison and Saratoga counties. Saratoga is remarkable for the number and variety of its springs, the waters of which are celebrated for their medicinal properties.

213. Among the most celebrated of the springs at Saratoga, are *Congress Spring, High Rock Spring, Hamilton Spring, Putnam's Spring, Iodine Spring, Pavilion Spring,* and *Empire Spring,* not any two of which are alike in the composition of their waters. Congress-Spring is the general favorite of visitors; it is highly acidulous. Hamilton, Putnam, and Pavilion springs are chalybeate. Iodine Spring contains but a small proportion of iron, and may be drank by a certain class of invalids, with whom iron proves a decided injury.

214. Virginia abounds with mineral springs, but the best known are the White and Blue Sulphur Springs of Greenbrier County, the Salt and Red Sulphur, and the Sweet in Monroe County, Hot and Warm in Bath, Berkley in Morgan, Fauquier White Sulphur in Fauquier, Shannondale in Frederick, Alum in Rockbridge, Jordon's White Sulphur in Frederick, Red in Alleghany, Grayson in Carroll, Botetourt in Roanoke, Holston in Scott, Augusta Springs and Daggu, Springs in Botetourt.

LESSON III.

RIVERS.

LODORE WATERFALL.

215. RIVERS commonly take their rise in springs, lakes, or the thawed ice and snow of elevated mountains. The smaller streams which unite to form them are variously called rills, rivulets, and brooks. Though rivers frequently rise in lakes, or spring from small elevations, the great storehouses from which the mightiest streams are fed are the ice-clad mountains of table-lands.

216. The *basin* of a river comprehends the entire country drained by it and its tributaries. The elevated land separating one basin from another is called the *water-*parting or *watershed,* of which the ridge of a house-roof affords an illustration. This is sometimes a lofty range of mountains, as the Alps, streams from which flow into opposite and distant seas. But commonly a water-parting has no great elevation, a slight rise of the surface being sufficient to separate streams whose mouths are thousands of miles apart. South America presents the rare example of two river-basins connected by a navigable natural channel. This is caused by the river Orinoco branching off in its upper course and proceeding by different channels to the sea.

217. South America is remarkable for containing the largest river-basin in the world, that of the Amazon measuring 2,275,000 square miles. Next to it, in point of extent, is the basin of the Mississippi and Missouri, in North America, measuring 1,333,000 sq. miles. The greatest river-basin in Asia is that of the Obi, 1,250,000 square miles; the greatest in Africa is that of the Nile, supposed to contain 1,240,000 square miles; and the greatest in Europe is that of the Volga, 528,000 square miles.

218. The *course* of rivers is commonly winding. In some cases it is so tortuous that its actual length is three times that of a straight line drawn from its source to its mouth, as the Forth in Scotland; in others, as the Hudson,—it is mainly straight. The Mississippi River is remarkable for its windings, or *bends,* as they are locally called.

WINDINGS OF THE MISSISSIPPI.

219. In 1848 the Raccourci Cut-off, an artificial channel, was made in this stream a little below the confluence of the Red River. The distance, about 30 miles, in the old channel a common steamboat would be three hours in going round, and now one can run through the cut-off in ten minutes up, and in two or three minutes down. The passage was at first a narrow one, but it has become so widened and deepened by the force of the current that the largest trees of the forest will go down root foremost, and the tops of them will disappear.

220. The actual meanderings of the Missouri and the Mississippi, or the difference between the direct distance from the source of the former to the mouth of the latter and that by their channels, is estimated at 2,148 geographical miles. The meanderings of the other principal streams, including that of their tributaries, is given approximately in the following table:

Rivers.	Meandering.
Rhine	240 geographical miles.
Elbe	840 " "
Rhone	352 " "
Dnieper	532 " "
Don	552 " "

Questions.—211. Other kinds of mineral springs? 212. Mineral springs of the U. States? What noted springs in the State of New York? For what is Saratoga remarkable? 213. Which are the most celebrated springs at Saratoga? 214. Principal springs in Virginia? 215. Where do rivers commonly take their rise? What are the great storehouses from which they are fed? 216. What is the basin of a river? The water parting or watershed? What elevated water-parting is mentioned? Have water-partings commonly a great elevation?

Questions.—What rare example is mentioned of two river basins being connected by a navigable natural channel? 217. For what is South America remarkable? Extent of the basin of the Amazon? Of the Mississippi and Missouri? The greatest river-basin in Asia, and its extent? In Africa, and its supposed ex ent? In Europe, and its extent? 218. What is said of the course of rivers? The Forth in Scotland? The Hudson? The Mississippi? 219. Raccourci Cut-off? 220. Meanderings of the Mississippi? Other rivers?

of rapids, falls suddenly to a depth of 153 feet. The cataract is divided by Goat Island into two parts. The largest of these, on the Canadian side, called the Horse-shoe Fall, from its shape, is 1,800 feet—more than one third of a mile—broad, and 153 feet in height. The fall, on the American side, is 600 feet in breadth and 164 feet in height. A cloud of mist points out the locality of the cataract, and under favorable circumstances its roar may be heard, it is said, at a distance of 40 miles.

FALLS OF ST ANTHONY.

226. The Falls of St. Anthony, in the Mississippi, about 400 miles from its source, and just above the mouth of the St. Peters, stand nearly at the head of the navigation of that stream. They have a descent of from 17 to 20 feet, and are thus described by a recent traveler: "Above the falls the river is about 600 yards in width. In its descent it is divided by Cataract Island, a high rocky mass, covered with trees and shrubbery. All around this island, above and below, are strewn huge masses of limestone rocks, heaped in Titanic confusion, and

attesting the mightiness of the water with which they seem to be contending."

227. Among the other noted falls in America are those of the Missouri, 500 miles from its source, a succession of rapids and cataracts, 26, 47, and 87 feet in perpendicular height; the rapids of St. Lawrence, above Montreal, extending about 9 miles; the Great and Little Falls of the Potomac, in Maryland; the Falls of Montmorenci, near Quebec, which descend 240 feet in an unbroken sheet; and the Cataract of Te-

FALLS OF MONTMORENCI.

quendama, in the environs of Santa Fé de Bogota, a magnificent fall of 574 feet.

Questions.—223. What is a rapid? A cataract? Cascade? 224. Where are the greatest cataracts? 225. What is said of the Falls f Niagara? Describe it. 226. Give particulars of the Falls of St. Anthony. 227. Other noted falls in America?

228. The Great Falls of the Missouri are the grandest in all North America, those of Niagara excepted; and although the latter exceed the former with respect to volume of water, depth of descent, and awful grandeur, yet the former are far more diversified and beautiful. These falls are within sixty geographical miles of the easternmost range of the Rocky Mountains. They were discovered by Captain Lewis while engaged in exploring the Missouri. Hearing a sound like that of a fall of water, he proceeded in the direction of it. "As he advanced, a spray arose above the plain like a column of smoke, and vanished in an instant. Toward this point he directed his steps; and having traveled seven miles after first hearing the sound, he reached the falls about mid-day. The hills as he approached were difficult of access, and 200 feet high. Down these he hurried with impatience, and seating himself on some rocks under the center of the falls, enjoyed the sublime spectacle of this stupendous cataract, which since the creation has been lavishing its magnificence on the desert, unknown to civilization."

229. The most noted falls of the Eastern Continent are Terni,

REICHENBACH.

Italy, a fall of the Velino, of 300 feet, usually regarded as the finest European cataract; Staubach, near Lauterbrunnen, in Switzerland, a perpendicular descent of 800 feet; Rhinefall, near Schaffhausen, a fall of the river Rhine in three distinct branches over a precipice upward of 80 feet high; Reichenbach, canton of Bern, a series of six falls, amounting to 1,000 feet; Ache, a small river of Bavaria, descending in five falls 2,000 feet; Evanson, a torrent of Mount Rosa, 1,100 feet; Cauvery, southern India, two grand falls near Seringapatam 810 feet; Girsupah, near a town of that name in the western Ghauts, a single fall of a considerable stream, 872 feet; the cataracts of the Nile, in Nubia and southern Egypt, of ancient celebrity; and Victoria Falls, on the Zambezi, which are described as scarcely less grand than Niagara.

LESSON V.

RIVERS—(CONTINUED).

230. THE *termination* of rivers is commonly by a single mouth, as in the instances of the St. Lawrence and the Hudson; but many streams, flowing through alluvial regions, carry along

Questions.—228. The Great Falls of the Missouri? Give an account of its discovery. 229. Give the particulars of the following falls and rapids: Terni, Staubach, Rhinefall, Reichenbach, Ache, Evanson, Cauvery, Girsupah, Nile, Victoria. 230. Termination of rivers? Sedimentary accumulations? Deltas? 231. Delta of the Mississippi? The rate of its formation? 232. Proportion of the sedimentary matter? Amount annually brought down? 233. Sedimentary matter in the Ganges?

sedimentary accumulations which they deposit at their mouths, producing formations of land, and dividing the main stream into branches. Such formations are called *deltas*, from their correspondence to the shape of the Greek letter Δ (delta).

DELTA OF THE MISSISSIPPI.

231. The delta of the Mississippi consists of a long, narrow tongue of land protruding for 50 miles into the Gulf of Mexico, at the end of which are numerous channels of discharge. This extensive formation has been produced by the deposit of the sedimentary matter brought down by that mighty stream. The rate of permanent advance of the new land has been very slow, not exceeding, it is supposed, one mile in a century.*

232. It has been calculated that the mean annual amount of solid matter which the water of the Mississippi contains is about $\frac{1}{1515}$ of its weight, and that it is equal to about $\frac{1}{2900}$ of its volume. It has also been estimated that the quantity of solid matter annually brought down by the river is 3,702,758,400 cubic feet, an amount which would cover over eleven miles square one foot deep.

233. The waters of the Ganges contain a much greater proportion of sedimentary matter, particularly during the season of the rains, which lasts about four months. The average quantity of solid matter suspended in the water during such times was by weight $\frac{1}{100}$th part; and the solid matter discharged is $\frac{1}{157}$th part in bulk, or 577 cubic feet per second. The total annual amount of mud discharged is estimated at 6,368,077,440 cubic feet, a mass equal in weight and bulk, it has been calculated, to eighty-two of the great pyramids of Egypt. The Hoang-Ho, which traverses the great alluvial plain of China, is supposed to bring down in one hour 2,000,000 cubic feet of earth, which so thickens and colors the ocean by its discharges as to originate the name of the Yellow Sea.

234. Rivers are grouped into two grand classes, the oceanic and the continental. *Oceanic* rivers are those which flow direct into the ocean, or into waters communicating with it. They form four distinct systems, belonging respectively to the Arctic, Atlantic, Pacific, and Indian oceans.

I. SYSTEM OF THE ARCTIC OCEAN.

Rivers.	Course.	Termination.	Length in English miles.
Lena	Eastern Siberia	Arctic Ocean	2,400
Olenek	Northern Siberia	Ditto	1,150
Yenesei	Central Siberia	Ditto	2,900
Obi	Western Siberia	Gulf of Obi	2,530

Questions.—Total annual amount? The Hoang-Ho? 234. Rivers, how grouped? Oceanic rivers? Systems? Name the principal rivers belonging to the system of the Arctic Ocean. Atlantic Ocean. Pacific Ocean. Indian Ocean.

* Sir Charles Lyell.

Rivers.	Course.	Termination.	Length in English miles.
Petchora	Northern Russia	Arctic Ocean	695
Dwina	Ditto	White Sea	760
Mackenzie	British America	Arctic Ocean	2,160

II. System of the Atlantic Ocean.

Rivers.	Course.	Termination.	Length
Neva	Northern Russia	Gulf of Finland	46
Vistula	Poland, Prussia	Baltic Sea	630
Elbe	Germany	German Ocean	690
Rhine	Switzerland and Germany	Ditto	760
Loire	France	Bay of Biscay	570
Rhone	Switzerland, France	Mediterranean	450
Danube	Austrian Empire, Turkey	Black Sea	1,630
Dnieper	Southern Russia	Ditto	1,200
Don	Ditto	Ditto	1,100
Nile (Blue N.)	Abyssinia, Nubia, Egypt	Mediterranean	2,600
Senegal	Western Africa	Atlantic Ocean	900
Niger	Ditto	Gulf of Guinea	2,300
Gariep	Southern Africa	Atlantic Ocean	1,000
Saskatchewan	British America	Hudson Bay	1,920
St. Lawrence	Ditto	Atlantic Ocean	2,070
Mississippi-Missouri	Central North America	Gulf of Mexico	4,100
Rio del Norte	Mexico	Ditto	1,400
Magdalena	New Granada	Caribbean Sea	800
Orinoco	Venezuela	Atlantic Ocean	1,200
Amazon	Brazil	Ditto	8,900
Araguay	Ditto	Ditto	1,300
St. Francisco	Ditto	Ditto	1,500
Plata	Ditto	Ditto	2,350

III. System of the Pacific Ocean.

Rivers.	Course.	Termination.	Length
Amour	Eastern Asia	Sea of Okhotsk	2,800
Hoang-Ho	China	Yellow Sea	2,600
Yang-tse-Kiang	Ditto	Ditto	8,200
Si-Kiang	Ditto	China Sea	1,110
Meinam	Siam	Gulf of Siam	900
Cambodia	Tibet, Cochin China	China Sea	2,000
Columbia	Oregon	Pacific Ocean	760
Colorado	Utah, New Mexico	Gulf of California	900

IV. System of the Indian Ocean.

Rivers.	Course.	Termination.	Length
Euphrates	Western Asia	Persian Gulf	1,000
Tigris	Ditto	Ditto	980
Indus	Northern India	Arabian Sea	1,700
Ganges	Ditto	Bay of Bengal	1,400
Brahmapootra	Tibet, Assam	Ditto	2,000
Irawaddy	Tibet, Birman Empire	Ditto	1,200
Murray	South Australia	Encounter Bay	1,000

235. *Continental* rivers are those which are confined exclusively to the continents, and discharge themselves into completely insulated lakes, or are lost in sands, deserts, or swamps. The principal continental rivers are stated in the following table:

Rivers.	Course.	Termination.
Volga	European Russia	Caspian.
Kour	Georgia	Ditto.
Terek	Circassia	Ditto.
Ural	European and Asiatic Russia	Ditto.
Amoor Jihon (ancient Oxus)	Turkestan	Lake Aral.
Sir or Sihoon	Ditto	Ditto.
Helmund	Afghanistan	Lake Zurrah.
Yarkand	Central Asia	Lake Lopnor.
	Many streams in Central Asia terminate in insulated lakes, or are lost in desert sands.	

Rivers.	Course.	Termination.
Jordan	Palestine	Dead Sea.
	Streams north of the African Sahara lost in sands.	
Rio Grande	Mexico	Lake Parras.
Desaguadero	Bolivia	Lakes and swamps.
Humboldt	Utah Territory	Lake.
Bear	Ditto	Great Salt Lake.
	Many other streams in Utah Territory are lost in sands, marshes, or lakes.	

236. The magnitude of rivers depends upon the area of their basins, the rain-producing character of the climate, and the arrangement of the surrounding mountains with reference to the prevailing winds. Lieut. Maury thus accounts for the vast rivers of the South American continent: "The northeast and southeast trade-winds, as they come across the Atlantic, filled with moisture, go full charged into the interior, dropping it in showers as they go, until they reach the snow-capped summits of the Andes, where the last drop, which that very low temperature can wring from them, is deposited to melt and feed the sources of the Amazon and the La Plata, with their tributaries."

237. The proportional quantity of water discharged by some of the principal rivers has been estimated as follows:

Thames	1	Don	38	Obi	179
Rhine	13	Volga	80	Nile	250
Loire	10	Euphrates	60	St. Lawrence	112
Po	6	Indus	188	Mississippi	338
Elbe	8	Ganges	148	Plata	490
Vistula	12	Yang-tse-Kiang	258	Amazon	1,280
Danube	65	Amour	160		
Dnieper	86	Lena	125		

238. Many rivers are subject to periodical inundations. It is to the annual overflowings of the Nile that Egypt owes her fertility. The rise commences about the 21st of June, attains its greatest height near the middle of September, and gradually diminishes to the time of the winter solstice. Both the Mississippi and the Missouri have annual floods during the spring months. Among the other streams subject to overflow are the Orinoco, Amazon, Euphrates, Tigris, etc.

VIEW ON THE NILE.

239. The following beautiful extract relates to the historic associations of rivers: "The rivers of the old world, while subordinate to those of the new in point of magnitude, have a historic and sacred interest in many instances wholly wanting to the latter. The Danube recalls the struggle of the ancient civilization with overwhelming barbaric invasions; the Tiber revives the story of Roman greatness; the Nile associates itself with the colossal power of the Pharaohs; the Tigris and Euphrates are connected

with the mighty dynasties of Assyria and Babylon, the captivity of God's chosen people, and that momentous hour when the hand of retributive justice unfolded the doom of monarch and of nation at a profane festivity of the royal house ; while the Jordan is imperishably linked with far more stupendous transactions : the miracle which divided its waters at the ingress of Israel into the land of promise ; and that voice from heaven which proclaimed the character of the Messiah at his baptism in the stream, placing his right to the universal homage of man, supported by the design of his mission, on the firm ground of his Divine nature : 'This is my beloved Son, in whom I am well pleased.'''[O]

LESSON VI.

RIVER SYSTEMS OF THE WESTERN CONTINENT.

240. IN no portion of the world are rivers found upon so magnificent a scale as on the continent of North and South America. The magnitude of these streams is owing to a variety of causes, some of which have been described ; as the situation and direction of the great mountain ranges, the extent of the river-basins, etc.: others will be explained in a subsequent part, which treats of wind, rain, etc.

241. NORTH AMERICAN RIVERS.—The continent of North America may, with reference to its drainage, be divided into four grand divisions: (1) The Atlantic slope, or that part which is drained by the rivers tributary to the Atlantic Ocean ; (2) the Valley of the Mississippi, lying between the Rocky and Alleghany mountains ; (3) the region to the west of the Rocky Mountains, the streams of which are tributary to the Pacific Ocean ; (4) the northern slope drained by streams tributary to Hudson Bay and the Arctic Ocean.

242. The Mississippi is the largest of the North American rivers ; it waters the southern half of the great plain, and brings to the sea the drainage of upward of a million of square miles. It rises in the small lake of Itasca, at an elevation of only 1,400 feet above the level of the sea, and flows in a southerly direction to its termination in the Gulf of Mexico, after a course of 2,400 miles. The Mississippi is navigable for 2,000 miles to the Falls of St. Anthony, a distance of about 400 miles below its source.

VIEW ON THE OHIO.

243. The Missouri has its origin in the Rocky Mountains, and runs for 2,500 miles in a southeasterly direction before it joins the Mississippi, making a total distance, from its source to its entrance into the Gulf of Mexico, of more than 4,000 miles. It is navigable from the point where it unites with the Mississippi up to the base of the Rocky Mountains, where the Great Falls occur (in latitude 46° 50').

244. During the period of the annual floods, large tracts of the adjoining country are overflowed, and immense damage done to property. Artificial embankments, called levees, are constructed along the lower course of the Mississippi to prevent inundation ; but in seasons of high floods they are often broken through by the force of the waters, forming what are known as crevasses. About one fifth of the whole surface of the State of Louisiana is periodically overflowed.

245. The Ohio is, next to the Missouri, the most important of the tributaries of the Mississippi ; it is formed by the junction of the Monongahela and the Alleghany rivers, which unite their waters at Pittsburg, where the river becomes navigable. The Ohio, about 950 miles in length, and from three quarters of a mile to upward of three quarters of a mile in breadth, flows through one of the most fertile portions of the Mississippi Valley. It has no fall, except a rocky rapid of 22½ feet descent at Louisville, around which a canal has been constructed sufficiently capacious to admit steamboats, though not of the largest class.

246. Among the other principal streams in the Valley of the Mississippi are the following : the St. Peters, Iowa, Des Moines, St. Croix, Wisconsin, and Illinois rivers, tributaries to the Mississippi above, and the Ohio, Arkansas, and Red rivers below, the junction of the Missouri ; the Yellow Stone, Platte, Kansas, and Osage rivers, tributaries to the Missouri ; and the Scioto, Miami, Wabash, Kanawha, Licking, Kentucky, Green, Cumberland, and Tennessee rivers, tributaries to the Ohio.

247. The St. Lawrence is the second great river of the North American continent, and by far the largest of the streams draining the Atlantic slope. Its channel from Lake Ontario to the sea is about 750 miles long, but (including the great chain of the lakes from which it derives its waters) the whole length, from the mouth of the St. Lawrence to the western extremity of Lake Superior, is 1,880 miles. Above Montreal, rapids occur which interrupt its navigation. During four months of the year the navigation is completely stopped by the ice.

248. Of the other streams draining the Atlantic slope, the principal are the Connecticut (400 miles), the Hudson (325 miles), the Delaware (300 miles), the Susquehanna (450 miles), the Potomac (400 miles), the James (450 miles), the Roanoke (350 miles), the Pedee (350 miles), the Santee (350 miles), the Savannah (400 miles), and the Altamaha (400 miles),—all of which flow into the Atlantic Ocean. The Appalachicola (600 miles) and the Mobile, farther to the westward,

Questions.—240. Magnitude of the American rivers ? To what owing ? 241. How may the continent of North America be divided with reference to its drainage ? First division ? Second ? Third ? Fourth ? 242. The Mississippi ? Its source, direction, and termination ? How far navigable ? 243. The Missouri ? How far navigable ?

Questions.—244. Floods ? Artificial embankments ? 245. The Ohio ? Its length, etc. ? Falls ? 246. Tributaries to the Mississippi above the junction of the Missouri ? Below the junction of the Missouri ? Tributaries to the Missouri ? To the Ohio ? 247. The St. Lawrence ? Its length from Lake Ontario ? Its whole length, including the lakes ? Rapids ? Ice ? 248. Other streams draining the Atlantic slope ? Their navigability ? The Hudson, how far navigable for the largest ships ? For steamers ?

* Rev. Thomas Milner.

low into the Gulf of Mexico. Most of these rivers are navigable for a considerable distance inland, and the Hudson (at the
mouth of which stands the city of New York) can be ascended
by the largest merchant ships 120 miles, and steamers nearly
60 miles.

249. The principal rivers draining the region to the west
of the Rocky Mountains—forming the third division—are the
Fraser (600 miles), Oregon or Columbus (750 miles), Sacramento (420 miles), and the Colorado (840 miles). Columbia,
the most considerable of these, receives several important tribu-
aries, the largest of which is the Lewis. The river Sacramento
waters the northern half of the valley lying between the Sierra
Nevada and the coast range of California, and forms at its
mouth the harbor of San Francisco: immediately above this
outlet it is joined by the San Joaquin, which drains the south-
rn half of the valley, and carries off the waters of Lake Tulare.
The streams tributary to the Sacramento and San Joaquin,
which flow in from the west, drain the gold regions of Cali-
ornia.

250. Of the streams draining the northern slope of North
America two divisions may be made,—one embracing the Nel-
on, Churchill, and other rivers flowing into Hudson Bay; and
he other, the Mackenzie, Coppermine, and other streams flow-
ng into the Arctic Ocean. The most considerable of these
treams is the Mackenzie, which issues from the western ex-
remity of the Great Slave Lake. During the summer it pours
, vast flood of waters into the sea, but is obstructed by ice
luring nine months of the year.

251. The Rio del Norte is a considerable stream of 1,400
miles in length, which rises in the southern part of the Rocky
Mountains and runs southwesterly into the Gulf of Mexico. To
he south of this stream are few rivers of any considerable
length. The San Juan, in Central America, possesses consid-
rable interest from its having afforded, since the discovery of
old in California, the principal channel of communication be-

tween the opposite shores of the Atlantic and Pacific oceans.
It has its origin in the lake of Nicaragua, on the southern portion of the plateau of Guatemala, and after a course of 120
miles empties itself into the Caribbean Sea.

LESSON VII.

RIVER SYSTEMS OF THE WESTERN CONTINENT—(CONTINUED).

252. SOUTH AMERICAN RIVERS.—The three principal rivers
of South America are the Orinoco, the Amazon, and the La
Plata, which drain respectively the northern, middle, and southern portions of the great plain.

253. The *Orinoco* (about 1,200 miles long) rises in the mountains of Guiana. About 130 miles below its source it sends off
to the southward a branch called the Casiquiare (200 miles in
length), which joins the river Negro, a tributary of the Amazon,
and thus effects a natural communication (navigable for boats)
between the basins of these two great rivers.

254. The principal tributaries of the Orinoco are the Guav-
iare, Meta, and Apure (rising in the Andes), on its left bank;
the Ventuari, Caura, and Caroni, on the right. Below the
village of Angostura, 280 miles from the mouth of the Orinoco,
no impediments to its navigation occur: above this its course
is occasionally interrupted by rapids, but in general it presents
a deep and navigable channel nearly to its source.

255. The *Amazon* (called the Maranon in the upper portion
of its course) rises in the small lake of Lauriococha, upon the
table-land of Pasco, amid the highest parts of the Peruvian
Andes. Thence it flows in a northerly course until it leaves
the mountains, and afterward crosses the great plain in an
easterly direction to the Atlantic Ocean. The whole length
of the Amazon is a little short of 3,900 miles.

256. The tributaries of the Amazon are numerous and on a
grand scale, some of them being equal to the largest
streams of the Old World. The principal are the
Napo, Putumayo or Ica, Japura, and Negro, on the
left bank; the Huallaja, Ucayali, Javary, Jutay, Jura
or Hyaruha, Teffé, Purus, Madeira, Tapajos, Xingu or
Chingua, and Tocantins, upon the right. The Madeira has a course of more than 2,000 miles before it
joins the Amazon.

257. At the distance of 700 miles below its source
the Amazon has a width of 800 yards, and during the
last 450 miles of its course it is nowhere less than four
miles in width, and at its mouth the channel is fifty
miles across. A recent exploration of this river* has
proved that it is navigable for vessels of the largest
class, during a considerable part of the year, from its
mouth to the very foot of the Andes, a distance of
about 3,500 miles. So great is the amount of water
which it brings down that its freshness is perceptible at a distance of more than 500 miles from
the coast.

FALLS OF THE MADEIRA.

Questions.—249. The principal rivers draining the third division? The Columbia?
Sacramento? San Joaquin? Gold regions, how drained? 250. What two divisions of the
reams draining the northern slope? The Mackenzie? 251. The Rio del Norte? San
Juan? 252. The three principal rivers of South America? 253. The Orinoco? Casiquiare?
54. The principal tributaries of the Orinoco? Its navigability? 255. The Amazon?

Questions.—Its course? Length? 256. What is said of the tributaries of the Amazon?
The principal tributaries on the left bank? Upon the right bank? The Madeira? 257.
Width of the Amazon? How far navigable? Illustrate the amount of water it brings down.

* By Lieut. Herndon.

258. The *Rio de la Plata* is a broad fresh-water estuary, which receives the waters of the Parana and Uruguay. The Parana has its rise in the mountains of Brazil, and, reckoning from its most distant source to the mouth of the Rio de la Plata, is 2,350 miles in length. About 760 miles above the sea it receives the Paraguay, which has a course of about 1,260 miles. The Uruguay (1,000 miles long) flows nearly parallel to the lower course of the Parana. Both the Parana and Paraguay are navigable for vessels of considerable draft to a distance of nearly 1,000 miles.

259. Among the other principal rivers of South America are the Magdalena (860 miles), which flows into the Caribbean Sea, and the Atrató (300 miles), into the Gulf of Darien. The Essequibo, Demerara, Berbice, Corentyn, Surinam, Maroni, and Oyapok flow into the Atlantic to the eastward of the mouth of the Orinoco. To the east and south of the Amazon are the Maranhao, Parnahiba, St. Francisco (1,500 miles), and many others belonging to the Atlantic coast. The Colorado (600 miles) and the Negro (800 miles) flow into the Atlantic southwest of the Rio de la Plata.

260. The rivers on the western coast of South America are very short; among the longest are the Biobio, in Chile, about 150 miles, and a few streams of nearly equal length to the northward of the Gulf of Guayaquil.

LESSON VIII.

RIVER SYSTEMS OF THE EASTERN CONTINENT.

261. EUROPEAN RIVERS.—Europe is divided into two principal river systems,—one embracing those streams which have a southerly direction, and flow into the Mediterranean, Caspian, and Black seas; the other comprising those which have a northerly course, and terminate in the North, Baltic, and White seas, and the Arctic Ocean.

262. In the west, these two systems are for the most part divided by the Alps and German mountains; but in the eastern part of Europe the division of the waters is merely a ridge of the great plain slightly elevated above the general level. This watershed begins on the northern declivity of the Carpathian Mountains, about the 23d meridian, in a low range of hills running between the sources of the Dnieper and the tributaries of the Vistula, from whence it winds in a tortuous course along the plain to the Valdai table-land, which is its highest point, 1,200 feet above the sea. It then declines northward toward the Onega, about the 60th parallel, and lastly turns in a very serpentine line to the Ural Mountains, near the 62d degree of north latitude.

263. The *Volga*, which is the largest river of Europe, rises in the plateau of Valdai, at the height of 1,100 feet above the sea; its entire course is 2,200 miles in length, and the area of its basin about 520,000 square miles, or nearly one seventh of the whole surface of the continent. The Volga is navigable through nearly its whole length, and has considerable depth. During the greater part of winter it is frozen over, but at other times is the highway of a most extensive traffic.

264. The *Danube*, 1,630 miles, is the second of the rivers of Europe, both in length of course and in the area of its basin, which is 310,000 square miles in extent. It rises in the Black Forest, to the north of Switzerland, at a height of 2,200 feet above the level of the sea, and flows in an easterly direction across the plateau of Bavaria, through Austria, the plain of Hungary, and between Bulgaria and Wallachia, until it enters the Black Sea.

265. The Danube is navigable from its mouth up to Ulm (10° east longitude), or throughout nearly the whole length of its course, excepting between the towns of Moldova and Gladova, where it passes, for a space of 60 miles, through a succession of rapids and shallows. The principal tributaries are the Inn, the Drave, the Save, and the Morava, on the south (or right-hand) bank; and the March, the Theiss, the Aluta, and the Pruth, on the north side. All of these are rivers of considerable magnitude.

266. The other principal streams which flow into the Black Sea are the Dniester (1,200 miles), the Dnieper (1,200 miles), and the Don (1,100 miles). The Dnieper is navigable from Smolensk to the sea, excepting for a distance of about 150 miles below Kiev, within which its navigation is impeded by rocks and cataracts.

267. The *Rhine* rises in the Alps, at an elevation of 6,580 feet, and flows in the upper portion of its course through Lake Constance. It has a length of 760 miles, and its basin is 70,000

BASLE, ON THE RHINE.

square miles in area: at Basle (where it is 765 feet above the level of the sea) its breadth is 800 feet, at Mentz about 1,700, and at Cologne 1,400. It is navigable for boats as high up as the Falls of Schaffhausen, a short distance below Lake Constance, and for vessels of some magnitude as high as Strasburg. The current is generally rapid, flowing at the rate of four or five miles an hour. Its principal affluents are the

Questions.—258. The Rio de la Plata? The Parana? Paraguay? Uruguay? How far are the Parana and Paraguay navigable? 259. Other principal rivers of South America? Rivers flowing into the Atlantic to the east of the mouth of the Orinoco? To the east and south of the Amazon? To the south of the Rio de la Plata? The rivers of the western coast? Bi-bio? 261. Into what two river systems is Europe divided? 262. Division of the two systems in the western part?

Questions.—In the eastern part? Describe the course of the watershed through eastern Europe. 263. Give particulars of the Volga. How far navigable? 264. Give particulars of the Danube. Its source, direction, and termination? 265. Navigation of the Danube? Tributaries? 266. Other streams which flow into the Black Sea? The Dnieper? 267. The Rhine? Its length, extent of its basin, etc.? Its principal affluents?

Neckar and Mayne on the right bank, and the Aar and Moselle on the left.

268. Of the other considerable rivers flowing into the Atlantic Ocean are—the Elbe (690 miles) and the Weser (380 miles) to the north, and the Meuse (550 miles), Seine (430 miles), Loire (570 miles), and Garonne (350 miles), to the south. Spain is watered by numerous rivers, as the Minho (200 miles), the Douro (460 miles), the Tagus (510 miles), and the Guadiana (450 miles) ; they are all navigable in the lower parts of their courses. The Guadalquiver (290 miles) is navigable for large vessels up to Seville.

269. The rivers which flow into the Mediterranean have generally short courses, owing to the nearness of the mountains on the north. The Ebro (420 miles) flows from the eastern side of the Spanish table-land. The Rhone (490 miles) rises in the highest region of the Alps, and passing in its course through Lake Geneva, below which it is navigable, falls into the Gulf of Lyons. The Arno (150 miles) and the Tiber (210 miles) both water the western side of the Italian peninsula ; the Po (450 miles) and the Adige (250 miles) flow through the plain of Lombardy, and enter the Adriatic Sea near its northern extremity.

270. Of the rivers flowing into the Baltic Sea are—the Duna (450 miles), the Niemen (400 miles), the Vistula (630 miles), and the Oder (550 miles). The Duna, the Vistula, and the Oder are navigable for the greater part of their courses. The Neva, which flows into the head of the Gulf of Finland, though only 46 miles in length, is of considerable importance, and has a vast volume of water, since it is the outlet of the great lakes of Ladoga and Onega : it has a mean breadth of 1,500 feet and a depth of 50 feet, but is frozen over for five months of the year.

271. The White Sea and Arctic Ocean receive several important streams, among which are the Dwina (760 miles), the Mezen (480 miles), and the Petchora (500 miles).

Eastern hemisphere. Of the other considerable rivers in the north of Asia are the Yenesei (2,900 miles) and the Lena (2,400 miles). The Obi, Yenesei, and Lena all rise in the mountains of the Altai system, and flow through the Siberian plain. Owing to the severity of the climate their waters are frozen during a great part of the year, and they are hence of little use for the purposes of navigation.

274. The Ganges (1,460 miles) and the Indus (1,700 miles), two of the most important rivers of Asia, both water the northern portion of Hindoostan. The Ganges, whose basin extends from east to west to the south of the Himalaya Mountains, flows in an easterly direction into the head of the Bay of Bengal. At its mouth it divides into numerous arms, which inclose a delta of immense extent (page 30) : its most western arm, called the Hoogly, upon which Calcutta is built, is the only one that is usually navigated. The Ganges is remarkable for the great extent of its fall : it is ascended by steamers as high as Allahabad, more than 800 miles from its mouth.

275. The Indus rises on the plateau of Tibet, to the northward of the Himalaya Mountains, at an elevation of more than 15,000 feet, and flows into the Arabian Sea. About 470 miles above its mouth the Indus receives on its left bank the river Chenaub, which collects the waters of the five streams of the Jeloam, the Chenaub, the Ravee, the Bayas, and the Sutlej. The district watered by these five rivers is called the Punjaub.* All the chief tributaries of the rivers, as well as the main stream, are navigable through nearly their entire length ; steamboats of considerable size can ascend to more than 500 miles distance from the sea, and smaller vessels to 1,000.

276. Of the other principal rivers belonging to the basin of the Indian Ocean are the Saleun and the Irawaddy (1,200 miles), both flowing into the Gulf of Martaban ; the Brahmapootra, the Godavery, the Krishna, and the Cauvery, into the Bay of Bengal ; the Nerbudda into the Gulf of Cambay ; and the united Euphrates (1,600 miles) and Tigris (980 miles) into the Persian Gulf.

LESSON IX.

RIVER SYSTEMS OF THE EASTERN CONTINENT—(CONTINUED).

272. ASIATIC RIVERS.—The rivers of Asia may, like those of Europe, be divided into two principal classes, separated by the mountains and table-lands which extend east and west through the interior. The northern division embraces the rivers which flow into the Arctic Ocean, and those (in the west) which terminate in inland seas or lakes unconnected with the ocean. The other and more numerous class includes the streams which have their origin in the mountains of the interior, and flow either southerly into the Indian Ocean, or eastwardly into the Pacific.

273. The Obi, which flows into the Arctic Ocean, is 2,530 miles in length ; its river-basin is 1,250,000 square miles in extent, being probably the largest basin of any river in the

RIVER TIGRIS, AT BAGDAD.

277. The seas to the east of Asia receive several large rivers, among which are the Amour (2,300 miles), which flows into the Gulf of Tartary ; the Hoang-Ho (2,600 miles), and the Yang-tse-Kiang (3,200 miles), both flowing into the Yellow Sea ; and the Cambodia (2,000 miles), into the Gulf of Siam.

* Properly, Peenj-ab, or five rivers.

278. The drainage of a large part of the Asiatic continent—probably not less than four and a half millions of square miles—is unconnected with any of the surrounding oceans, but is received into inland seas or lakes, of which the Caspian and Aral possess the most extended basins. The principal continental rivers of Asia are the Kour (550 miles), the Amoo or Jihon (1,300 miles), and the Sihoon (1,150 miles), flowing into the Sea of Aral. The Tarim or Erghue (900 miles) flows into the Lake of Lop, in the center of the continent. The Helmund (600 miles), which rises in the plateau of Afghanistan, falls into Lake Zurrah; and the Jordan, in Palestine, into the Dead Sea.

279. AFRICAN RIVERS.—The *Nile* is the most considerable river of Africa: it carries off the waters from the northern and western sides of the plateau and mountains of Abyssinia, and discharges itself into the Mediterranean. The Nile is formed by the junction (in latitude 15° 40′ north) of two streams,—the Bahr-el-Azrek (Blue River), and the Bahr-el-Abiad (or White River): the latter is generally admitted to constitute the main channel of the river. The wonderful secret of the source of the Nile, for centuries a geographical mystery, has lately been unraveled by the intrepid and successful travelers, Speke, Burton, and Grant. It rises in an extensive lake called Victoria Nyanza, lying beneath the equator. A short distance of the course of the river remains unexplored; but its origin is no longer doubtful.

280. Though the Nile has so great a length of course—probably not much short of 3,500 miles—its basin is of very limited extent. For a distance of 1,400 miles above its mouth it receives no tributary. Through the middle and lower portion of its course, the Nile flows in a narrow valley inclosed on either side by steep rocks; the width of this valley varies from one to two miles in Nubia and Upper Egypt to as many as ten or twelve miles lower down the stream.

281. The *Niger* (or Quorra) is the largest of the African rivers which flows into the Atlantic Ocean. It rises in the mountains of Soudan, where the main tributary is known as the Joliba, pursues a northeasterly course to the neighborhood of Timbuctoo, thence a southeasterly course, passing through a wide opening of the Kong Mountains, and enters the eastern extremity of the Gulf of Guinea by several mouths. The length of the Niger is perhaps about 2,300 miles; above the place of its passage through the Kong Mountains it receives the waters of the Chadda, a broad and deep tributary. The Niger has been ascended by a steamboat to more than two hundred miles above the junction of the Chadda, but the extreme unhealthiness of the climate, in the district through which its lower course lies, has contributed to the failure of many attempts made to explore this part of Africa, and to establish commercial relations with the inhabitants.

282. Besides the Niger, the principal rivers upon the west coasts of Africa are the Senegal, the Gambia, the Rio Grande, the Rokelle, the Volta, the Zaire or Congo, the Coanza, and the Gariep or Orange. Both the Senegal (900 miles) and the Gambia (650 miles) are navigable rivers; they flow in a westerly direction, and draw their waters from the same mountain ranges in which are the sources of the Niger. The Gariep or Orange River, in the southern part of Africa, has a length of upward of 1,000 miles. The principal river upon the eastern side of Africa is the Zambesi, which brings down a great volume of water, and is said to be navigable for boats through a distance of more than 900 miles. The Lufiji, Juba, and many other rivers of the eastern coast, have not been explored, and are but little known.

LESSON X.

LAKES.

283. FRESH-WATER lakes occur in the greatest numbers, and upon the largest scale, in the northern regions of the globe. Nearly all the lakes of any considerable extent in North America are situated north of the 40th parallel, while in Europe and Asia, the regions peculiarly characterized by fresh-water collections are, for the most part, north of the 50th parallel. Salt-water lakes have a more southerly distribution, and are very abundant in eastern Europe, and central and southern Asia.

284. Lakes may be divided into four classes, according to certain physical peculiarities. The *first* class includes those which have no outlet, and do not receive any running water. Lake Albano, near Rome, is an example. Many of these lakes are situated in elevated districts, and are generally small; it has been supposed that they are the craters of extinct volcanoes, and are supplied by springs.

285. The *second* class comprises those which receive water, but have no apparent outlets. The Caspian Sea and Lake Aral belong to this division. The Caspian is about 600 miles long; its extreme breadth is 300 miles, though its average breadth is not more than 100 miles. This most remarkable lake receives the waters of the Volga, a river which has a course of 2,200 miles, and brings down more than 518,000,000 cubic feet of water every hour. The Ural and many other streams of considerable magnitude are also received by the Caspian; but its level is not changed, though it has no perceptible outlet by which to discharge the water it receives.

286. Lake Aral presents the same phenomenon, and, though not to be compared in extent to the Caspian, receives two large rivers, the Sihoon and Amoo or Jihon. The difficulty in explaining the nature of these lakes is to account for the constancy of their level, which might be expected to rise considerably, as they are daily receiving so large a body of water. The opinion was entertained that they are connected by some internal channel with the sea, and it was supported by the fact that the water of both the Caspian and Lake Aral is salt, and

Questions.—278. Extent of the surface drained into the inland seas? Principal continental rivers of Asia? Into what seas or lakes do they respectively flow? 279. What is said of the Nile? How formed? Source of the Nile? In what does it rise? Exploration of the river? 280. Length? Basin? Width of the valley in different parts? 281. The Niger? Its source, direction, and termination? Length? How far ascended by a steamboat? The climate of the district in which its lower course lies? 282. Other principal rivers upon the west coasts of Africa?

Questions.—The Senegal and the Gambia? The Gariep? The principal river upon the eastern side? Other rivers? 283. Where are fresh-water lakes most abundant? In North America? In Europe and Asia? 284. Into how many classes may lakes be divided? First class? Example. Situation of these lakes, etc.? 285. Second class? Example. The Caspian? What waters are tributary to it? 286. Lake Aral? Opinion formerly entertained? How supported? How is this hypothesis disproved? How may the constant level be accounted for? What other lakes belong to this class?

contains marine productions; but it has been ascertained that the Caspian is not less than 84 feet below the level of the Black Sea, thus completely disproving the hypothesis that they have a connection. It is thought that the phenomena referred to may be accounted for by evaporation and filtration. Besides the Caspian Sea and Lake Aral, there are numerous other bodies of water of this class, the receptacles of the continental rivers. (See table of Continental Rivers, page 31.)

287. A *third* class comprehends all those lakes which receive no streams, but give birth to some. Many of these lakes occupy very elevated situations, and are the sources of some of the largest rivers. They are no doubt supplied by springs, the waters of which rise in their reservoirs until its level is sufficiently high to admit a discharge. The lake in Monte Rotondo, in Corsica, is one of this class, and is situated 9,000 feet above the level of the sea.

288. The *fourth* class includes all those lakes which both receive and discharge water, being by far the most numerous division. They commonly receive the waters of many rivers, and have but one outlet. The origin of such rivers is easily explained. Should a hollow present itself in the course of a river, it is evident that it must be filled to the level of some part of its banks before the river can proceed, and this would produce a lake. But it may happen that there is a general declivity from various parts of a district toward some central valley, and then the waters of a number of rivers may be brought into it, while at the same time the continuation gives but one course by which the waters can be discharged. A description of some of the largest lakes of this class will be given in the next lesson.

289. Most lakes occur at varying elevations above the level of the sea, while some are much below it. The highest known lake in the world is Sir-i-kol, in Asia. It is the source of the

LAKE TITICACA.

Amoo River, and is 15,630 feet above the level of the ocean. Lake Titicaca, in Bolivia, has an elevation of 12,785 feet; Tzana or Dembea, in Abyssinia, 6,076 feet; Lake Baikal, in Asia, 1,793 feet; Constance, 1,299 feet; Geneva, 1,229 feet; Great Salt Lake, in Utah Territory, 4,200 feet; Superior, 623 feet; Huron and Michigan, 591 feet; Erie, 565 feet; Ontario,

234 feet. The Caspian Sea, Lake of Tiberias, and the Dead Sea are each below the sea-level,—the first 84 feet, the second 600 feet, and the third 1,317 feet.[*]

LESSON XI.

LAKES—(CONTINUED).

290. NORTH AMERICAN LAKES.—The largest lakes in North America are Superior, Huron, Michigan, Erie, and Ontario, which are connected with the sea by the channel of the river St. Lawrence; Winnipeg, which is drained by the river Nelson into the Hudson Bay, and the Athabasca, Great Slave, and Great Bear, which empty their waters into streams tributary to the Arctic Ocean.

291. *Lake Superior* is the largest fresh-water formation on the globe, computed to have an area of 40,000 square miles; length 420 miles; extreme breadth, 165; height above the level of the Atlantic, 623 feet; greatest depth, 1,200 feet. There is reason to believe, from the appearance of the shores, that the waters of this, as well as the other Canadian lakes, formerly occupied a much higher level than they reach at present. The amount of water carried off by its outlet, the river of St. Mary, is much less than that received by its tributaries, from which circumstance it is inferred that the evaporation from its surface must be very great.

292. *Lake Huron*, remarkable for its brilliant transparency, has an area of 25,000 square miles. It is about 240 miles in length, from 180 to 220 in breadth, and is 591 feet above the level of the sea. The outline of this lake is very irregular, and its shores are described as consisting of clay cliffs, rolled stones, abrupt rocks, and wooded steeps. The greatest depth of Lake Huron is found to be nowhere more than 450 feet. *Lake Michigan*, which lies wholly within the United States, is connected with Lake Huron by means of the navigable channel Mackinaw. It is about 300 miles long, and has an area of about 25,000 square miles.

293. *Lake Erie* has an area of about 11,000 square miles; its surface is 565 feet above the sea. This lake is said to be the only one in the whole Canadian chain in which there is any perceptible current, a circumstance which is supposed to be attributable to its comparative shallowness, its average depth being not more than 60 or 80 feet. The current of Lake Erie, which runs always in one direction, combined with the great prevalence of westerly winds, and the occurrence of sunken reefs and

---

Questions.—287. Third class? What is said of many of the lakes of this class? How supplied? Monte Rotondo? 288. Fourth class? How is the origin of such lakes explained? 289. Varying elevations and depressions of lakes? Highest known lake, its elevation, etc.? Give the elevations of the other lakes mentioned. State the depressions of the Caspian, Lake of Tiberias, and the Dead Sea. 290. Which are the largest lakes in North America, and how drained?

Questions.—291. Give particulars of Lake Superior. Change of level? 292. Give particulars of Lake Huron. Lake Michigan. 293. Lake Erie. What is said of its current? Its navigability?

* According to the measurement made by Lieutenant Lynch, in 1848, the exact depression of the Dead Sea below the Mediterranean was found to be 1,316.7 feet.

rocky banks, form serious obstacles to the safe and easy navigation of this lake. The shallowness of the water of Lake Erie
likewise causes it to be more readily and more permanently
affected by frost, so that its navigation is usually obstructed by
ice for some weeks every winter, while that of the other lakes
continues open and unimpeded.

294. *Lake Ontario* has a computed area of 10,000 square
miles, 234 feet above the sea-level, and 331 feet below the level
of Lake Erie. Its depth is said to be very great, and it is
navigable throughout its whole extent for the largest ships. Its
outlet is a spacious channel studded with islands, collectively
denominated the Thousand Isles, but no less than 1,692 have
been actually counted.

295. *Lake Champlain* (about 500 square miles) belongs to
the same basin as the great lakes above described, and is connected with the St. Lawrence by the river Richelieu. *Lake
George*, noted for its picturesque scenery, and for the transparency of its waters, is situated west of the southern extremity

LAKE GEORGE.

of Lake Champlain, with which it is connected by a short stream.
It is about 30 miles long, and from one to two miles broad.

296. The *Great Salt Lake* (about 2,600 square miles), situated in the great basin between the Rocky Mountains and the
Sierra Nevada (111), is about 70 miles in length, from 30 to 35
miles in breadth, and is about 4,200 feet above the level of the
sea. Its waters are saturated with common salt, and when the
lake is low, considerable quantities of this substance are precipitated to the bottom of the lake, or, rather, are there crystal-
lized. No living animal can exist in this lake. It receives the
waters of the Bear, Weber, and other rivers, but, like other
lakes in this region, has no connection with the ocean.

297. Upon the Mexican plateau is the large lake of *Chapala*
(about 650 square miles), which is discharged into the Pacific
by the river Santiago. *Lake Nicaragua* (about 3,500 square
miles), in Central America, lies at an elevation of about 128 feet
above the sea. The distance between its western shore and the

coast of the Pacific is only eleven miles; it is drained by the
San Juan, which flows into the Caribbean Sea.

298. SOUTH AMERICAN LAKES.—South America has few
lakes of any great extent. The largest is *Lake Titicaca* (about
8,800 square miles), situated on a plateau of that name, at an
elevation of 12,785 feet, and surrounded by some of the highest
summits of the Andes. The water of Lake Titicaca is fresh; a
river called the Desaguadero, which leaves its southern extremity, flows into the smaller lake (or marsh) of Aullagas, or
Uros, which lies at 490 feet lower level, and the water of which
is salt.

299. *Lake Maracaybo* (5,000 square miles), near the coast
of the Caribbean Sea, is connected by a narrow strait with the
Gulf of Maracaybo, and has brackish water. The *Lake dos
Patos* and *Lake Mirim* are on the southeast coast of Brazil.

LESSON XII.

LAKES—(CONTINUED).

300. EUROPEAN LAKES.—There are two principal lake-regions in Europe, one lying around the Baltic, and situated
within its basin; and the other embracing the Alpine system
of mountains. The lakes situated in the former of these regions
possess, in general, greater magnitude, while the latter are distinguished by their great elevation above the sea, and by the
grandeur of the scenery among which they lie.

301. The following tables give the dimensions of the principal European lakes, together with their elevation and greatest
depth, where these particulars have been ascertained.

LAKES SITUATED ROUND THE BALTIC.

		Area in sq. miles.	Height.	Depth.
IN RUSSIA.....	Ladoga	6,330		
	Onega	3,280		
	Ilmen	390		
	Peipous or Tchoudskoé	1,250		
	Pskov	280		
	Bieloe	420		
	Saima	2,000		
	Enara°........................	1,200		
IN SWEDEN....	Wener	2,186	144	288
	Wetter	840	288	432
	Maelar	780	0	

LAKES BELONGING TO THE ALPINE SYSTEM.

		Area in sq. miles.	Height.	Depth.
IN SWITZERLAND..	Geneva....................	240	1,230	1,012
	Neufchatel...............	115	1,437	426
	Luserne..................	99	1,430	600
	Zurich...................	76	1,332	600
	Constance or Boden See	228	1,299	964
IN HUNGARY.....	Neusiedler See	150	850	13
	Balaton or Platten See.......	250	918	86
IN ITALY.....	Lago Maggiore.............	152	678	2,622
	Como....................	66	684	600
	Garda...................	183	320	

Questions.—294. Give particulars of Lake Ontario. The Thousand Isles. 295. Lake
Champlain. Lake George. 296. Great Salt Lake. What are its waters? 297. Give particulars of Lake Chapala. Lake Nicaragua. 298. What is said of South America? Lake Titicaca? 299. Lake Maracaybo? Other lakes? 300. How many lake-regions are there in
Europe, and where are they respectively situated? How are the lakes of each division
characterized? 301. What is the area of Lake Ladoga? Of Onega?

Questions.—Of other lakes in Russia? Give particulars of Lake Wener. Of other lakes
in Sweden. Of Geneva. Of other lakes in Switzerland. Of Lake Maggiore. Of other
lakes in Italy.

* The waters of Lake Enara, however, communicate with the Arctic Ocean, not with
the Baltic.

302. Lakes are very numerous in Scotland, especially in the middle and northern parts. They are mostly long and narrow bodies of water, occupying the deep hollows within the elevated mountain valleys. The largest lake in Scotland, and also in Great Britain, is *Loch Lomond* (45 square miles), which is 24 miles in length, and 7 miles in its greatest breadth.

303. Of the lakes in Ireland the largest is *Lough Neagh* (150 square miles), situated in the north of Ireland, and the *Lakes*

LAKE KILLARNEY.

of *Killarney* (three in number), noted for their beautiful scenery, in the south.

304. ASIATIC LAKES.—The largest fresh-water lake in Asia is Lake Baikal, situated among the northern offsets of the Altai mountain-system: it has an area of about 15,000 square miles, and lies at an elevation of 1,793 feet above the level of the sea. Its water is fresh, and abounds in fish. It is annually frozen over for a period of five or six months, and may be traversed on sledges.

305. Among the smaller lakes of Asia are Balkash, Upsa-nor, Zaizang, Issi-Kol, Bosteng, Lop, Koko-nor, Bouka-nor, and Tengri-nor—all on or adjacent to the high plateaus in the interior of the continent; Tung-ting and Poyang, in China; Zurrah and Bakhtegan (both salt), on the plateau of Afghanistan and Persia; Ooroomiah, Van, and Goukcha (the two former of which are salt), on the Armenian table-land; the salt lake of Koch-hissar, in Asia Minor; with Lake Tiberias and the Dead Sea in Palestine.

306. AFRICAN LAKES.—The largest body of inland water known in Africa, until recently, was *Lake Tsad*. It lies in the central part of the continent, and is several thousand square miles in area; but its waters are very shallow. It is not known to have an outlet, but is fresh, and clear, and probably has a channel of discharge like other fresh-water lakes.

307. Recent explorations in the southeast of Africa have made known the existence of several large lakes, the most extensive of which is *Lake Victoria Nyanza* or *Ukerewe*, the long-sought source of the Nile. Its area is not yet determined, but is believed to considerably exceed that of Lake Tsad. Its height is 3,553 feet above the level of the sea. Not far

from this, at a lower elevation, are *Lake Little Luta Nzige* (unexplored), and *Lake Tanganyika* or *Ujiji*. Among other noted African lakes are *Tzana* or *Dembea*, in Abyssinia, and *Nyanja* or *Nyassi*, and *Shirva*, in or near the borders of Mozambique.

LESSON XIII.

THE OCEAN.

308. THE vast body of water which surrounds the land and penetrates its coast is comprehended under the general name of the ocean. For convenience sake it is divided into five portions, named, respectively, the Arctic, Atlantic, Indian, Pacific, and Antarctic oceans. These, with their branches, are as follows:

		Branches.
I.		
ARCTIC OCEAN	Extends from the northern shores of America, Europe, Asia, and the arctic circle, around the north pole	Baffin Bay. White Sea. Gulf of Kara. Gulf of Obi.
II.		
ATLANTIC OCEAN	Bounded on the west by America; east by Europe and Africa; north by the arctic, and south by the antarctic circle—divided by the equator into the North and South Atlantic	Baltic with its gulfs. North Sea. Mediterranean. Black Sea. Hudson Bay. Gulf of Mexico. Caribbean Sea.
III.		
PACIFIC OCEAN	Inclosed between America on the east; Asia, the Sunda Isles, and Australia on the west; and the arctic circle on the north; the antarctic on the south—divided by the equator into the North and South Pacific	Sea of China. Yellow Sea. Sea of Japan. Sea of Okhotsk. Sea of Kamtchatka. Behring Strait. Gulf of California. Bay of Panama.
IV.		
INDIAN OCEAN	Bounded by Africa on the west; the Sunda Isles and Australia on the east; southern Asia on the north; and the antarctic circle on the south	Red Sea. Arabian Sea. Persian Gulf. Bengal Sea.
V.		
ANTARCTIC OCEAN	Extends from the antarctic circle around the south pole.	

309. The Arctic Ocean has been but partially explored. Various efforts have been made to reach its higher latitudes, but, up to this time, they have proved unsuccessful, in consequence of the impenetrable fields of ice which are met with, and the impossibility of remaining in those regions with safety for a great length of time. During the winter months the waters of the Arctic Ocean are covered with ice, which in summer is broken up and drifted into lower latitudes, where it is dissolved.

Questions.—302. What is said of the lakes in Scotland? Loch Lomond? 303. Lakes in Ireland? 304. Give particulars of Lake Baikal. 305. What other Asiatic lakes are mentioned? 306. Give particulars of Lake Tsad. 307. Of Lake Victoria Nyanza. Lakes near it? Other noted African lakes?

Questions.—308. Under what general name is the vast body of water which surrounds the land comprehended? How is it divided? Describe the situation, and mention the principal branches of the Arctic Ocean, Atlantic, Pacific, Indian, Antarctic. 309. What is said of the Arctic Ocean? Why have the efforts made to reach its higher latitudes been unsuccessful?

310. The floating masses of ice in the arctic waters are of two kinds, sheet-ice and icebergs, which have quite an independent origin. *Sheet-ice* is that which is formed by the freezing of the ocean's surface, and is generally level like that of lakes; it rises from two to eight feet out of the water. Vast fields, 20 or 30 miles in diameter, have been found in the Arctic Ocean; sometimes they extend 100 miles, so closely packed together that no opening is left between them. Smaller sheets are called *floes.* Fields and floes, when much broken up, the fragments crowding together, form what is called *pack-ice,* which, when much elongated, is called a *stream.* When the parts of a pack are loose and open, so that a vessel may sail between them, it is called *drift-ice.*

311. In 1850, Lieut. De Haven, commanding the Grinnell Expedition in search of Sir John Franklin, proceeded into the Arctic Ocean a considerable distance north of Wellington Channel. Here, in the early part of October, while drifting about among large masses of floating ice, his vessels (two in number) were frozen in so firmly that it was impossible, with all the means at command, to disengage them from the ice. In this state they were drifted back through Wellington Channel and Lancaster Sound into Baffin Bay, thence southeasterly through this bay to about latitude 66° north, where, after having been confined in the ice nearly eight months, and having drifted not less probably than 1,500 miles, they were liberated from their icy fetters.

312. *Icebergs* are fresh-water formations; and, towering like cliffs to a considerable height, they present a very different aspect from ice-fields. They are produced on the shores of arctic lands by the freezing of melted snow, like the glaciers of Switzerland. The frozen masses projecting into the sea, yield to its undermining and wrenching power, by which immense blocks are broken off, constituting icebergs. These huge masses are drifted southward 2,000 miles from the places of their origin to melt in the Atlantic, where they cool the

ICEBERGS.

water and air to a great distance around. Icebergs vary from a few yards to miles in circumference, and are often 1,000 feet high.

313. It is supposed that the point of the greatest cold is near the northern border of Little Grinnell Land, north of Welling-

ton Channel, and that not far to the north and west of it there is, in summer, a comparatively open sea, or "Polynia." The latter opinion is supported by the fact that beasts and fowls are known to migrate over the ice from the mouth of

SCENE IN THE ARCTIC OCEAN.

Mackenzie River and its neighboring shores to the north; and that, in the highest latitudes yet reached, both animal and vegetable life appear to be more abundant a few degrees farther south, and the waters exhibit a higher temperature. In further confirmation of this conjecture, Dr. Kane reports the discovery of a great body of open water north of Cape Constitution (latitude 81° 22′ north), on the coast of Greenland; along the shores of which the exploring party traveled for many miles, and which "was viewed from an elevation of 580 feet, still without a limit, moved by a heavy swell, free of ice, and dashing in surf against a rock-bound shore."

314. The Atlantic Ocean is the best known of any of the divisions of the great deep, it being the highway of the world's commerce, and constantly traversed by hosts of vessels in which millions of property and thousands of lives are embarked. It extends upward of 9,000 miles from north to south, with a width varying from little more than 900 miles between Norway and Greenland, to 1,700 miles between Cape St. Roque, in Brazil, and the coast of Sierra Leone, in Africa.

315. The North Atlantic, though generally very deep, is remarkable for immense shoals occurring in the North Sea, and to the southeast of Newfoundland. It is also noted for the immense portion of its surface occupied by sea-weed (*fucus natans*), closely matted together, forming what is sometimes known as the "Grassy Sea." A region of this weed extends along the meridian of 40° west longitude, and between the latitudes 20° and 45° north, bearing the name of "Banks of Fucus." It occurs thence in varying quantities to the Bahamas, the area occupied being equal to 1,000,000 geographical square miles and upward,—more than one third the extent of the whole territory of the United States.

316. The Pacific Ocean has about twice the area of the Atlantic, extending upward of 9,000 miles from north to south, and from east to west 12,000 miles. It was so called by the

Questions.—310. Of what two kinds are the floating masses of ice? Describe the sheet-ice. Its extent. Floes. Pack-ice. Streams. Drift-ice. 311. What part of the Arctic Ocean was reached by Lieut. De Haven? What subsequently happened to his vessels? How long were they confined and how far did they drift? 312. What are icebergs? How produced? How far south do they sometimes drift?

Questions.—313. Supposed point of the greatest cold? Open sea or "Polynia"? Confirmation of this conjecture? 314. The Atlantic Ocean? Its extent? 315. The North Atlantic? For what noted? Where are the Banks of Fucus situated? Where else does it occur? Extent of the sea so occupied? 316. What is said of the Pacific Ocean? Why so called? For what remarkable?

where it is of great depth. 6· The seasonal changes of the temperature of the air do not affect the ocean beyond the depth of 300 feet. 7. The greatest heat of the surface, 88° 5′ of Fahrenheit, is found in the Gulf of Mexico, and in one of the havens of New Guinea.

322. The *color* of the ocean is generally of a deep bluish green, but it varies with every gleam of sunshine or passing cloud, from the deepest indigo to green, and even to a slaty gray. It is different in different localities, depending upon local causes. It is white in the Gulf of Guinea and black around the Maldives. Between China and Japan it is yellowish, and west of the Canaries and Azores it is green. In some parts, as off California, it has a vermilion hue; in others, as the eastern division of the Mediterranean Sea, a purple tint prevails.

323. These various shades are, in most instances, caused by myriads of marine animalcules which pervade the deep; and the magnificent appearance, known as the phosphorescence of the ocean, is owing to the phosphorescent brilliancy of these microscopic tribes. The bed of the ocean, in shallow places, often imparts a tinge to the superincumbent waters, while the gray or turbid appearance, near the mouths of large rivers, arises from the sediment washed in from the land.

324. The *depth* of the ocean was, until recently, a subject of speculation only. The experiments made during the past few years have added more to our knowledge of the depth of the ocean and the shape of the oceanic basins, particularly that of the Atlantic, than was ever before known.

325. Like the dry land, the bottom of the sea is diversified with slopes, plains, table-lands, eminences abruptly projecting to within a few feet of the surface, or just peering above the waves, and with enormous depressions. It has been generally supposed that the depth of the sea is about equal to the height of the land, the lowest valleys of the ocean's bed corresponding with the summits of the loftiest mountains.

326. The experiments in deep-sea sounding initiated a few years since by Lieut. Maury, formerly of the United States National Observatory, will probably throw some light upon this conjecture. Before a systematic investigation was thus attempted, no well-directed efforts to fathom the lower abysses had been made. Navigators had tied weights to lines and thrown them overboard with the view of measuring the depth; but the lines were often unwieldy, and there was no certain means of knowing whether the plummet had reached the bottom, or, if it had reached the bottom, at what moment.

327. More recent investigations have led to the belief that there is in the ocean, as in the air, a system of circulation, which, by currents and counter currents, upper and under, keeps the waters perpetually in motion. For it has been found that, generally speaking, when a sounding is made in the deep sea, though the vessel from which it is made be perfectly at rest, and though it be known that the plummet has reached the bottom, yet the line will continue to run out, and unless it be suffered to run out, or the plummet be detached from it, a strain

so great is brought to bear that it breaks. It is the undertow, or a system of currents and counter currents below, which it is supposed produces this strain.

328. Most of the vessels of the navy are now furnished with twine made especially for deep-sea soundings; and the results already obtained have enabled the officers at the National Observatory to construct a map of the basin of the North Atlantic Ocean, which shows the depressions of the solid parts of the earth's crust below the sea-level, and which gives us, perhaps, as good an idea of the profile there as geographers have of the contrasts afforded by the elevations of the land in many parts of the earth.

329. The deepest soundings ever reported were made in the North and in the South Atlantic Ocean. Lieut. J. C. Walsh, commander United States schooner Taney, being furnished with a large quantity of iron wire made expressly for the purpose, obtained, on the 15th November, 1849, latitude 31° 59' north, longitude 58° 43' west, a cast of the plummet, when, after 34,200 feet had run out, the wire parted without reaching bottom, as it was thought. On the 12th of February, 1853, Lieut. Berryman, of the Dolphin, in latitude 32° 55' north, longitude 47° 58' west, obtained a cast of the lead, using the small twine as a sounding line. At this trial 39,600 feet ran out; when the line parted, and it was consequently thought that the plummet had not reached the bottom. On the 5th of April, 1852, latitude 36° south, longitude 44° 11' west, Lieut. Parker, of the United States frigate Congress, using a 32-lb. cannonball for his plummet, and sounding twine like that of Berryman's, made an experiment at deep-sea soundings, when 49,800 feet of line ran out before it parted. The time occupied for this sounding was eight hours and a quarter.

330. The next great sounding was made by Capt. Denham, of H. M. ship Herald, 30th October, 1852, latitude 36° 49' south, longitude 37° 06' west, with 46,236 feet. He had been furnished with sounding twine from the United States frigate Congress, and instead of a 32-lb. shot, his sinker was a 9 lb. lead. By the light which subsequent experience has thrown upon the subject of deep-sea soundings, all four of these immense depths have had their accuracy questioned, and it is believed with reason.

331. An instrument has been invented by Passed Midshipman J. M. Brooke, of the United States navy, which enables the officers who now attempt deep-sea soundings to detach the plummet from the line the moment it strikes the bottom, and then to haul up, attached to the line, specimens of the bottom. In this way specimens have been obtained from the depth of 12,000 feet (about 2¼ miles). These specimens have been examined with a microscope by Prof. Bailey at West Point, and found to consist entirely of minute sea-shells, not a particle of sand or gravel, or any foreign matter being among them. From this it is inferred that the water at the bottom of the sea is comparatively at rest.

332. The deepest part of the North Atlantic Ocean is prob-

ably a little to the south of the Grand Banks of Newfoundland. There is a place there somewhat in the shape of a boot, which none of the officers of the navy have so far been able to fathom. The deepest soundings that have been satisfactorily made show that, in all other parts, the North Atlantic Ocean is not more than 25,000 feet in depth. The soundings which have been made by the navy have established the fact that there is a plateau, or shelf, at the bottom of the ocean between Newfoundland and Ireland, quite shallow enough for the wires of a submarine telegraph, and quite deep enough to keep them beyond the reach of icebergs.

LESSON XV.

THE OCEAN—(CONTINUED).

333. The ocean is subject to a motion of three different kinds: it is agitated by the action of the wind, producing waves; by tides, which result from the attraction of the moon and sun; and by currents, produced under various circumstances, and resulting from a variety of causes.

334. Waves are produced by the action of the winds on the surface of the water, and vary in size from mere ripples to enormous billows. Their height in open sea depends upon the force and duration of the winds, and the angle at which it bears down upon the waters; but in lakes and bays it is affected by the shallowness of the waters and the character of the shores;

on which account the shallow waters of Lake Erie are more readily disturbed by winds than the deeper lakes, Ontario and Huron.

335. Waves are not, as appearances would indicate, an onward flow of water. This is proved from the fact that a floating body merely rises and falls with little or no progression. Waves agitate the water but a little way below the surface, and it is supposed that the effect of the strongest gales does not extend deeper than 200 feet.

Questions.—328. With what are most of the vessels of the navy provided? Results obtained? 329. When have the deepest soundings been made? Give particulars of the sounding made by Lieut. J. C. Walsh. By Lieut Berryman. By Lieut Parker. 330. By Capt. Denham. What is thought of these soundings by the light of subsequent experiments? 331. How have specimens of the bottom of the ocean been obtained? Of what have these specimens been found to consist?

Questions.—332. Where is the deepest part of the North Atlantic supposed to be? What do the soundings show with respect to the depth of the North Atlantic in all other parts? What other important fact has been established by these soundings? 333. What are the three different kinds of motion to which the ocean is subject? 334. How are waves produced, and how do they vary? Upon what does their height depend? How is the height affected in lakes and bays? 335. How is it proved that waves are not an onward flow of water?

336. The crest of a wave : (b, b) is the ridge or highest part,

and in strong winds is usually covered with foam ; the trough (c) is the depression between two waves, and is as much below as the crest is above the general level of the ocean. In estimating the elevation of a wave, the perpendicular height from the trough to the crest is taken.

337. Waves are sometimes said to run mountains high, but this is a popular exaggeration. The highest rise noticed in the Mediterranean is 16 feet, and 20 feet off Australia. During a storm in the Bay of Biscay, the highest waves measured scarcely 36 feet from the base to the summit. In the South Atlantic the result of several experiments gave only an entire height of 22 feet, and a velocity for the undulations of 89 miles per hour, the interval between each wave amounting to 1,910 feet. Off the Cape of Good Hope, notoriously the cape of storms, according to its former name, 40 feet is considered the extreme height of waves, or 20 feet above and below the general level of the ocean.*

338. The sea does not regain its placidity immediately after the subsidence of the winds which set it in motion, but continues to heave with mighty undulations for a considerable time afterward. This movement is called the "swell." It frequently occurs, that while the swell is advancing in one direction, the wind rises from an opposite quarter, producing a series of compound waves, and giving to the deep a very complex aspect.

339. *Tides* are those regular alternate risings and fallings of the waters of the ocean and of bays, rivers, etc., which communicate freely with it. They arise from the attractive influence of the sun and moon, the latter being the more powerful agent. The sea rises, or flows, as it is called, by degrees, about six hours ; it remains stationary about a quarter of an hour ; it then retires, or ebbs, during another six hours, to flow again after a brief repose. Thus high and low water occur twice every lunar day, or the period elapsing between the successive returns of the moon to the meridian of a place, which is 24 hours 50½ minutes.

340. The theory of the tides may be thus explained: Let E represent the earth surrounded by water in every part, and

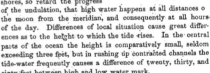

m the moon. As a solid body draws toward it any other body, by a force which varies with its distance from the point attracted, then the water at a will be more powerfully attracted than those at e and f, and the result will be a bulging out of the water at a, immediately next the lunar body.

341. But while high water is thus produced at a, it is also produced at the same time in the opposite hemisphere at d. Different causes have been assigned for this phenomenon, but it

is probably likewise owing to the attractive influence of the moon. The earth's center, E, will be more powerfully drawn toward the moon than the waters at d, and recede from them, producing the same effect as though they receded or rose up from the center of the earth. As the waters can not rise in one place without falling in another, they are depressed at e and f when elevated at a and d.

342. If the earth remained stationary upon its axis, there would be two elevations and depressions of the waters in each place in a month, or the time the moon is making her circuit round the earth. But by the diurnal rotation, the moon passes every day the meridian of every part of the globe, producing daily two seasons of high and low water.

343. The time of high tide does not coincide with the time when the moon is on the meridian of the place, a period of several hours ordinarily intervening between the time of the moon's transit and that of high water. The explanation of this is, that the water, having received motion, continues to rise after the moon has passed from its meridian, the impulse continuing for some time after the moon's transit.

344. Besides the attraction of the moon, the influence of the sun is concerned in elevating the waters of the ocean. The solar attraction is not so strong as the lunar, because, though a much larger body, the sun is at a far greater distance from the earth. The tidal influence of the sun has been calculated to be about one third that of the moon.

345. Sometimes the sun and moon act in conjunction, as at the seasons of new and full moon, a and d, and then the tides rise the highest, and are called *spring-tides ;* but when the moon is in quadrature, as at b and c, it acts in opposition to the sun, and then occurs the lowest, or *neap-tides.*

346. Though high tides occur in open seas soon after the moon has passed the meridian, yet in other places shoals and channels, peninsulas and capes, and the diversified form of shores, so retard the progress of the undulation, that high water happens at all distances o the moon from the meridian, and consequently at all hours of the day. Differences of local situation cause great differences as to the height to which the tide rises. In the central parts of the ocean the height is comparatively small, seldom exceeding three feet, but in rushing up contracted channels the tide-water frequently causes a difference of twenty, thirty, and sixty feet between high and low water mark.

347. The cradle of tides is supposed to be that part of the Pacific Ocean to the southeast of Australia. Proceeding from this quarter, a wave advances into the Indian Ocean, reaching the island of Sumatra, the southern point of Hindoostan, the

* Rev. Thomas Milner.

island of Madagascar, and the Cape of Good Hope about the same time. [See map No. 1.]

348. Entering the Atlantic, the wave proceeds northerly and westerly, bringing high water successively to all parts on the western shores of Africa and eastern shores of America. It moves with much greater rapidity through the central parts of the ocean than along the coast—in consequence of which it reaches the islands of Cuba and Newfoundland, and Cape Blanco, in Africa, simultaneously. The Atlantic coast of the United States receives the wave from the east, while the western coast of Europe receives it from the west, the direction, in the latter case, being nearly the reverse of what it had previously to doubling the Cape of Good Hope.

349. Along the shores of South America, from Rio Janeiro to the Falkland Islands, a wave advances from east to west, bringing high tide later on going southward, as if the wave came from the north. On the western coast of South America the tide travels from north to south, between the Gulf of Panama and the Strait of Magellan. And on the same side of North America, it travels from the Gulf of Panama northward to Queen Charlotte Sound.

350. The height to which tides rise varies greatly in different places. At St. Helena the rise does not exceed three feet; at St. Malo's, on the north coast of France, the spring tides rise 50 feet; at Chepstow, on the British Channel, 60 feet; in the Bay of Fundy, between Nova Scotia and New Brunswick, it is known to rise 70 feet. On some occasions the rapidity of the waters is so great as to overtake animals feeding on the shores.

TIDE TABLE FOR THE COAST OF THE UNITED STATES.*

LOCALITY.	Interval betw. time of moon's transit and time of high water. Mean. H. M.	Diff. betw. greatest and least H. M.	Rise and Fall. Mean. Feet.	Rise and Fall. Spring. Feet.	Rise and Fall. Neap. Feet.	Mean Duration. Flood. H. M.	Mean Duration. Ebb. H. M.	Mean Duration. Stand. H. M.
Portland	11 25	0 44	8.8	10.0	7.6	6 14	6 12	0 90
Boston	11 29	0 44	10 1	13.1	7.4	6 16	6 18	0 9
New Bedford Harbor	7 57	0 41	3.8	4.6	2.8	6 30	5 83	0 42
Newport	7 45	0 94	3.9	4.6	3.1	6 21	5 8	0 23
New Haven	11 16	1 8	5.8	6.6	5.1	6 94	6 5	
New York	8 13	0 46	4.8	5.4	3.4	6 0	6 25	0 28
Old Point Comfort	8 17	0 50	2.5			6 1	6 95	
Baltimore	18 83	0 43	1.8	1.5	0.9	5 54	6 83	
Smithville	7 19	0 47	4.5	5 5	3.8	6 1	6 96	0 96
Savannah	8 18	0 51	6.5	7.6	5.5	5 4	7 28	0 14
Key West	9 92	1 19	1.4	2.8	0.7	6 59	5 25	0 12

LESSON XVI.

THE OCEAN—(CONTINUED).

351. CURRENTS constitute the third oceanic movement. They consist of vast oceanic streams which keep up a perpetual circulation of the waters, transferring them from one hemisphere to another—from the Pacific to the Atlantic, and to the Pacific again—and from the polar seas to the warm regions of the torrid zone.

Questions.—348. Its progress after entering the Atlantic Ocean. What three points are reached simultaneously? 349. What is the direction of the wave south of Rio Janeiro? Describe its movements on the western coast of South America. On the western coast of North America. 350. Mention the height to which tides rise at different places.

* The author is indebted for the above table to the kindness of A. D. Bache, Superintendent of the United States Coast Survey.

352. Currents are due to a variety of causes; as the influence of tides and winds,—the evaporating power of the sun,—the expansion and contraction of water by heat and cold,—and the revolution of the earth upon its axis.

353. The effect of the rise and fall of tides in producing an alternate flowing of currents in opposite directions, is perceived in channels between islands, or between islands and the mainland. Thus, in the channel which connects Long Island Sound with the Harbor of New York, known as the East River, strong currents alternately prevail in opposite directions, as the tide ebbs or flows.

354. Evaporation by solar heat is another cause of oceanic currents. Large quantities of water raised from one tract of the ocean are transported to some other, where the vapor is

GIBRALTAR.

condensed and falls in the form of rain; this, in flowing back, to restore equilibrium, causes sensible currents. A perpetual stream flows into the Mediterranean from the Black Sea through the Bosphorus and the Hellespont, and another from the Atlantic through the Strait of Gibraltar. No counter, lateral, or submarine currents have been discovered sufficient to dispose of the quantity of water flowing inward; hence the inference that the inward current goes to supply the waste caused by an enormous evaporation.

355. The expansion and contraction of water by heat and cold are perhaps the principal causes to which currents are due. Warm water is specifically lighter than cold, and when certain portions become heated, they rise by reason of their buoyancy above the general surface, and are replaced by surrounding colder and heavier fluid flowing in beside or beneath them; while they flow off above.

356. The revolution of the earth upon its axis is still another powerful cause in producing currents, particularly those of the equatorial regions, which have commonly a westerly direction. The winds of tropical climates, which blow continuously, or during long periods in one direction, also lend their influence in affecting this class of oceanic movements.

357. Currents may be classed into constant, periodical, variable, counter, and drift currents. Constant currents are pro-

Questions.—351. Of what do currents consist? 352. To what are they due? 353. Where is the effect of the rise and fall of tides in producing currents perceived? Example. 354. How does evaporation by solar heat operate to produce currents? Mention what is said of the Mediterranean Sea. 355. What are perhaps the principal cause to which currents are due? Explain how they affect the currents. 356. What other causes are instrumental in producing currents? 357. How may currents be classed? What are constant currents? Periodical currents? Variable currents? Counter currents? Drift currents?

duced by the rotation of the earth, differences of temperature in the waters of the ocean, and other causes not yet fully understood. *Periodical* currents are principally due to the action of tides, though they are doubtless affected somewhat by the land and sea breezes and monsoons. *Variable* currents are occasioned by tides, winds, and the melting of ice in the polar regions. *Counter* currents are the streams that flow alongside or beneath, and in opposite directions to, other currents. *Drift* currents are the effect of permanent and prevailing winds upon the surface of the sea, and a variety of other causes.

358. EASTERN POLAR OR GREENLAND CURRENT.—The Eastern Polar or Greenland Current originates in the Arctic Ocean north of Asia. It sweeps around the northern shores of Spitzbergen and Iceland, and flows southwesterly between the latter island and Greenland. Near Cape Farewell it sends off a branch which runs into Baffin Bay, but its principal flow appears to be southwesterly between the Gulf Stream and the neighboring coast of America.

359. The breadth of this current is in some places from 250 to 300 miles. Its velocity varies, in different parts of its course, from eight or nine to fifteen or sixteen miles per day. The icy masses it bears along, and which are frequently swept around Cape Farewell, are supposed to be about two months in making the circuit to Baffin Bay and thence to the coast of Labrador.

360. This current is distinguished for the great amount of drift-wood which it floats along and casts upon the shores of Spitzbergen, Iceland, and other lands lying in its way. The masses of floating wood thrown upon the island of Jan Mayen often equal, it is said, the whole of the island in extent. It is supposed that this timber comes from the forests of Siberia, and is carried into the Arctic Ocean by the streams of northern Asia.

361. Recent observations in high northern latitudes show that the Eastern Polar or Greenland Current presents one of the most formidable difficulties in exploring the polar regions. Parry, who attempted to reach the north pole by means of boat-sledges and reindeer, traveled over the surface of the deep to nearly lat. 83°, which seemed to be the utmost limit of animal life. Here he found that when, according to his reckoning, he had traveled ten or eleven miles toward the north, he had actually gone four miles to the south, owing to the current. The success of the expedition was thus rendered hopeless. The two vessels of the Grinnell Expedition which were sent out, under the command of Lieut. De Haven, to search for Sir John Franklin, after having penetrated far into Wellington Channel, were inclosed firmly in the ice, and drifted backward through Baffin Bay, a distance of not less than fifteen hundred miles, thus defeating the plan of one of the noblest and most humane enterprises ever undertaken.

362. EQUATORIAL CURRENT.—The most extensive movement of the ocean is that which proceeds from east to west, on each side of the equator, and is therefore called the Equato-

rial Current. It originates in the general inflowing of waters from the temperate and polar regions, but especially from the Antarctic Ocean and neighboring seas. These waters, overbalancing the heated and therefore expanded, and specifically lighter waters at the equator, cause them to rise above the general level and, overflow in a constantly spreading stream. But as the *inflowing* waters come from parts of the earth's surface which lie nearer the poles or the earth's axis, and have a less rapid diurnal movement (from west to east) than the parts near the equator, they accordingly partake of this movement; and hence, as they approach the equator, fall behind the more swiftly advancing points of its surface, and thus acquire a western tendency. Accordingly the Equatorial Current, fed by these westerly moving waters, has a resulting westward course; and, crowding against the eastern shores of the continents and of the great islands which separate the Pacific from the Indian Ocean, divides into various streams, most of which flow off to the north and south.

363. In the Indian Archipelago and neighboring seas, however, it separates into numerous small branches that take different courses according to the channels through which they flow: hence the variable currents prevailing in the Indian Ocean, which render navigation so dangerous. A large volume of water forces its way through the islands, and joins the great equatorial current which moves on toward the eastern coast of Africa. The greater portion of the stream flows round north of the island of Madagascar, and sweeps through the channel of Mozambique, after which, being joined by other currents from the east, it moves toward the southern extremity of Africa, where it is said to unite with a current descending along the western coast of the grand division, and to flow thence to the Antarctic Ocean.

364. The Equatorial Current, on reaching Cape St. Roque, the most eastern point of South America, is separated into two branches. One proceeds southward along the coast of South America, under the name of the Brazilian Current, and terminates in a region of *variable* currents, the chief of which—observed as moving to the eastward with an increasing velocity—has been designated as the South Connecting Current.

365. The other and principal branch of the Equatorial Current of the Atlantic is known as the Guiana Current. It runs from off Cape St. Roque, across the mouth of the Amazon, and after skirting the low coast of Guiana, and passing through the Caribbean Sea, it enters the Gulf of Mexico between the island of Cuba and the peninsula of Yucatan.

366. MEXICAN GULF STREAM.—This is the most powerful current known, and the most important in consequence of the extent to which it affects the navigation of the Atlantic. It originates in the Gulf of Mexico, the waters of which are characterized by a remarkably high temperature. It pours forth at a rate of from three to five miles an hour through the straits of Florida, and flows in a northeasterly direction along the whole coast of the United States, expanding in volume

Questions. - 358. Origin of Eastern Polar or Greenland Current? Its course? Branch? Principal flow? 359. What is the breadth of this current? Its velocity? 360. For what is it distinguished? 361. What do recent observations in high northern latitudes show? Describe the attempt of Parry to reach the north pole. The drift of the two vessels of the Grinnell Expedition. 362. Which is the most extensive movement of the ocean, and what is it called? How does it originate? Cause of this movement of the waters? How do the inflowing acquire a peculiar tendency?

Questions.—Result concerning the Equatorial Current? 363. What happens on its reaching the Indian Archipelago? Describe its subsequent course. 364. Its branches in the Atlantic? What of its southern branch? 365. What is the name of the other branch, and where does it run? 366. What is said of the Mexican Gulf Stream? Where does it originate? Its velocity through the straits of Florida? Its subsequent course? Its course after striking the banks of Newfoundland? Great whirlpool of the Atlantic?

and diminishing in rapidity.* On striking the banks of Newfoundland, it sets to the east: the northern portion, however, sweeps toward Iceland, Norway, and the British Isles; the southern portion flows to the Azores, where it turns south and enters the Equatorial Current on the coast of Africa, and is conducted again to the west, to re-enter into itself in the Gulf of Mexico. Thus the waters of the Atlantic Ocean between the parallels of 11° and 48° constitute a whirlpool of prodigious extent, by which a single particle of water describes a circuit of over 11,000 miles in the space of two years and ten months.

367. The Gulf Stream, as it issues from the straits of Florida, is a dark indigo-blue; the line of junction between it and the green waters of the Atlantic is plainly seen for hundreds of miles. This line is finally lost to the eye as the stream goes north, though it is preserved to the thermometer for several thousand miles. From observations made with the deep-sea thermometer, it has been ascertained that "the stream, as far as the banks of Newfoundland, flows through a *bed* of cold water, which cold water performs to the warm the office of *banks* to a river."†

368. Coming from the heated caldron of the Gulf of Mexico, the waters of the Gulf Stream have a high temperature, which is gradually lost as they reach higher latitudes. "The maximum temperature of the Gulf Stream is 86°, or about 9° above the ocean temperature due the latitude. Increasing its latitude 10°, it loses 2° of temperature. And, after having run 3,000 miles toward the north, it still preserves, even in winter, the heat of summer. With this temperature it crosses the 40th degree of north latitude, and there, overflowing its *liquid banks*, it spreads itself out for thousands of square leagues over the cold waters around, and covers the ocean with a mantle of warmth that serves so much to mitigate in Europe the rigors of winter. Moving now more slowly, but dispensing its genial influences more freely, it finally meets the British Islands. By these it is divided, one part going into the polar basin of Spitzbergen, the other entering the Bay of Biscay, but each with a warmth considerably above ocean temperature. Such an immense volume of heated water can not fail to carry with it beyond the seas a mild and moist atmosphere. And this it is which so much softens the climate there."‡

Questions.—367. What is the color of the Gulf Stream? What fact has been ascertained respecting it? 368. What is said of its temperature?

Questions.—What is its maximum or greatest temperature, and how many degrees is it above that due the latitude? Describe how the Gulf Stream serves to moderate the climate of Europe.

---

* Different opinions have been formed respecting the cause of the Gulf Stream. It is supposed by some, that the waters of the Mexican Gulf have a higher level than those of the Atlantic in consequence of the trade-winds and the influx of the Equatorial Current; and that the current is merely the *running off* of the water in order to restore an equilibrium. Accordingly, the stream has been likened to "an immense river descending from a higher level into a plain." But Lieut. Maury has very satisfactorily disproved this theory, and shown that, "instead of *descending*, its bed (the bed of the stream) represents the surface of an inclined plane from the north, up which the lower depths of the stream *must* ascend." It is safe to assume, respecting the cause of this remarkable current, that it is influenced much by the excessive temperature imparted to the waters of the Gulf of Mexico. The course of the Gulf Stream has been assigned to the differ-

ence in density between the waters of the Caribbean Sea and the Gulf of Mexico, and those of the Baltic and the North seas. The waters of the former contain a larger proportion of salt, and are consequently heavier than sea water; while those of the latter, being only slightly impregnated with saline matter, are much lighter than common sea water. This difference in density destroys the equilibrium and produces a current; "for wherever equilibrium be destroyed, it is restored by motion, and motion among fluid particles gives rise to currents, which, in turn, constitute circulation." It is more probable, however, that its curvilinear direction is due to the less rapid eastward movement of the earth's surface as it approaches the higher latitudes.

† Lieut. M. F. Maury.   ‡ *Ibid.*

PART III.

THE ATMOSPHERE.

LESSON I.

COMPOSITION OF AIR.

A TMOSPHERE is the name of that thin, transparent, and highly elastic fluid which surrounds the earth on every side, and accompanies it in its diurnal revolution upon its axis and its annual motion round the sun. It is lighter than either land or water, and rises above them, but is kept by the force of gravity close to the surface of the earth, where its use is indispensable to all living creatures. It is the medium through which sound, light, and odor are transmitted; it is the vehicle in which moisture is raised and diffused; and the agent by which that diversity of color so pleasing to the eye is produced in natural objects.

370. Atmosphere is unlike the great divisions of land and water in not being perceptible to the touch unless in agitation. Its existence as a material substance is evident the moment it is set in motion. It not only carries away in its progress the lighter substances with which it comes in contact, but, when greatly agitated, uproots trees, crumbles rocks, and overturns buildings. Its motion is applied as a mechanical force, and as such is of vast use to man in wafting his vessels over the ocean.

371. The atmosphere is composed principally of two different gases, termed oxygen and nitrogen, the relative proportions being 21 parts of the former to 79 of the latter. It contains a small but variable proportion of aqueous vapor, and a still smaller proportion of carbonic acid gas. The proportions of oxygen and nitrogen are definite, but the amount of aqueous vapor fluctuates. Under ordinary circumstances, the composition of 1000 parts of the atmosphere may be stated as follows:

Oxygen	210.0
Nitrogen	775.0
Aqueous vapor	14.2
Carbonic acid	0.8
	1094.0

372. The same proportions of oxygen and nitrogen are found in the atmosphere of all countries, and at all elevations, over land and over sea, on the summit of the highest mountains and at their base, at the equator, and in high northern and southern latitudes. The quantity of carbonic acid gas is, however, greater near the level of the sea in summer than in winter; greater during the night than the day; and rather more abundant on the summit of high mountains than on plains.

373. Oxygen and nitrogen are extremely different in their properties. Oxygen gas is a supporter of combustion, and is required for the support of animal life, while nitrogen, in its unmixed state, is destructive to both. Without oxygen, fires would cease to burn, and all animals would immediately expire. By the process of breathing it is taken into the lungs and goes to purify the blood. When the blood is brought into the lungs it is of a dark purple color, but it then throws off the hydrogen and carbon, and receives oxygen, which gives it a bright red color. A portion of the nitrogen that is received by the lungs appears to be absorbed, while the other and larger part is rejected and thrown back again into the atmosphere in which it immediately rises, being lighter than air.

LESSON II.

PROPERTIES OF THE ATMOSPHERE.

374. THE general properties of the atmosphere are transparency, fluidity, weight, and elasticity. *Transparency* is that

state or property it possesses by which it suffers rays of light to pass through it, so that objects can be distinctly seen through it. The various degrees of clearness in the atmosphere are owing to particles of vapor and other substances which float in it. Distant objects sometimes appear twice as near as at others, a phenomenon occasioned by the difference in the purity of the atmosphere, or its freedom from aqueous and other particles.

375. By the *fluidity* of the atmosphere is meant that quality it possesses which renders it impressible to the slightest force, and by which the particles easily move or change their relative positions. Fluidity is a property common to liquid and aeriform substances. The atmosphere, like other fluids, presses in all directions, upward as well as downward, and is capable of supporting light bodies.

Questions.—369. Of what is atmosphere the name? What is said of its lightness? What else is remarked of it? 370. How is atmosphere unlike the great divisions of land and water? Its effects when in motion? Its motion how applied? 371. Of what is the atmosphere principally composed, and in what proportions? What other substances does it contain? State the composition of 1000 parts of the atmosphere. 372. What is said of the invariable proportions of oxygen and nitrogen in common air? In what localities, and at what times, is the quantity of carbonic acid greater?

Questions.—373. What is said of oxygen gas? Of nitrogen gas? What would happen without oxygen? What is the color of the blood when brought into the lungs? What change then takes place? What becomes of the nitrogen received into the lungs? 374. What are the general properties of the atmosphere? What is transparency? To what are the various degrees of clearness owing? Why do distant objects appear sometimes twice as near as at others? 375. What is meant by the fluidity of the atmosphere? How does the atmosphere press?

376. The air is ponderable, or has *weight*. The pressure or weight exerted upon every square inch of the earth's surface is equal to about 15 pounds. In consequence of its fluidity it presses equally in every direction, and the human body, of the ordinary size (supposed to measure 15 square feet), sustains the enormous pressure of 31,536 pounds or 14 tons. We do not feel the pressure, owing to its acting uniformly on all sides, and because the air within our bodies perfectly counterpoises the external pressure.

377. The weight of a column of the entire atmosphere is equal to that of a column of water of the same base about 34 feet high, or a similar column of mercury 30 inches high. The pressure diminishes as we ascend, according to a scale, which is nearly certain. From numerous observations it has been ascertained that

	3½		2	
	7		4	
	10½		8	
at the	14	miles above the level	16	times lighter than at
height of	17½	of the sea, the air is	32	the earth's surface.
	21		64	
	24½		128	
	28		256	

378. The pressure of the atmosphere is indicated by the barometer (the measure of weight), an instrument consisting of a column of mercury poised or pressed up into a vacuum by the weight of the atmosphere. The mercury rises or falls according to the pressure of the atmosphere, its range, at the level of the sea, being from about 28 to 31 inches.

379. The barometer is used for determining the height of mountains. At the level of the sea the pressure is greatest in consequence of the weight of all the superincumbent atmosphere, and hence at that point the highest column of mercury will be sustained; but as we ascend, this superincumbent pressure is diminished, and consequently the mercury falls. Thus, Humboldt, at the foot of Mount Chimborazo, found the barometer to stand exactly at 30 inches; but on ascending the mountain to the elevation of 19,000 feet, it was very little higher than 14 inches. In the pass of Antarangra (one of the highest passes of the Andes), Lieut. Herndon found the barometer to stand at 16.73 inches, indicating an elevation of 16,044 feet. Experiments have proved that the mercury will fall about $\frac{1}{10}$th of an inch for every 100 feet of perpendicular height, or one inch for every 1,000 feet.

380. The density or pressure of the atmosphere is, by another method, made subservient to the measurement of heights; namely, by observing the boiling point of water, which decreases in a ratio nearly equivalent to the decrease of atmospheric pressure. At the level of the sea, water boils, or passes into the state of steam, at 212° Fahrenheit, but at the Hospice of the Great St. Bernard it boils at the lower temperature of 203°, and on the top of Mont Blanc at 186°. In the pass of Antarangra, Lieut. Herndon found water to boil at 182°.5. From these and other observations, it may be inferred that a difference of one degree in the boiling point of water, as indi-

cated by the common thermometer, answers very nearly to 550 feet of elevation.

381. The *elasticity* of the atmosphere is the property it possesses of occupying less space under the influence of certain forces, and returning to its original volume when the influence is withdrawn. Hence its density is not uniform, but, as before explained, diminishes from below upward. The height of the atmosphere is not known, but it is supposed to extend to about fifty miles. By far the greater portion of it is within fifteen or twenty miles of the earth's surface; and at a much less distance it becomes so rarefied as to be incapable of supporting life.

382. Travelers on high mountains have experienced sensible, and sometimes painful, proofs of the rarefied state of the air. In very elevated regions the thinness of the air diminishes the intensity of sound, renders breathing difficult, and produces a loss of physical strength. The blood burst from the ears and lips of Humboldt in attempting to reach a high elevation in the Andes. He experienced the same difficulty in kindling and maintaining a fire at great heights, which Marco Polo felt on the mountains of central Asia. In the high regions of the Andes a malady prevails, called *veta*, which is thus spoken of Lieut. Herndon: "Veta is the sickness caused by the rarity of the atmosphere at these great elevations. The Indians call it veta or vein, because they believe it is caused by veins of metal diffusing around a poisonous infection. The affection displays itself in a violent headache, with the veins of the head swollen and turgid, a difficulty of respiration, and cold extremities."

LESSON III.

WINDS.

383. The atmosphere remains at rest so long as its density is unchanged; but as soon as the equilibrium is broken by any cause whatever, a motion occurs, which is called *wind*. If, in one part of the atmosphere, the air becomes dense, it passes away to those parts where the density is less, in the same manner as air compressed in a pair of bellows escapes by the orifice. We may compare this displacement of air to that of water in rivers; it is a flowing of the aerial ocean from one region toward another.

384. The currents of the atmosphere perform many invaluable services to man. They renew the air of cities; and they mitigate the climates of the north by bringing to them the heat of the south. They transport the clouds from the sea to the interior of continents, thus aiding to fertilize regions which would otherwise become arid and uninhabitable. They waft the sails of the navigator around the globe, bring distant nations into familiarity, and are thus greatly instrumental in the diffusion of civilization and Christianity throughout the world.

Questions.—376. What is the pressure or weight on every square inch of the earth's surface? What pressure does the human body of the ordinary size sustain? Why do we not feel the pressure? 377. To what is the weight of a column of atmosphere equal? 378. By what instrument is the pressure of the atmosphere indicated? What is the range of the mercurial column at the level of the sea? 379. For what is the barometer used? Where will the highest column of mercury be sustained, and why? Why does it fall as we ascend? What did Humboldt discover? Lieut. Herndon? What have experiments proved? 380. By what other method may the density of the atmosphere be made subservient to the measurement of heights?

Questions.—At what degree of Fahrenheit does water boil at the level of the sea? At the Hospice of the Great St. Bernard? On the top of Mount Blanc? In the pass of Antarangra? What elevation is found to answer the difference of one degree in the boiling point? 381. What is the elasticity of the atmosphere? How does its density diminish? What is the supposed height of the atmosphere? 382. What have travelers on high mountains experienced? What effects are produced by the thinness of the air? 383. How long does the atmosphere remain at rest? What occurs when this equilibrium is broken? If in one part of the atmosphere the air becomes dense, what follows? To what may this displacement of the air be compared? 384. Describe the uses of the atmospheric currents.

385. To indicate the direction of the wind the horizon is divided into eight equal parts, and the wind is designated by giving it the name of the points of the horizon whence it blows. The eight kinds of winds are *north, northeast, east, southeast, south, southwest, west,* and *northwest.* It is customary to write merely the initial of these words, that is: N., N. E., E., S. E., S., S. W., W., N. W.

386. The general direction of the wind near the surface of the earth is indicated by vanes. They are commonly placed on elevated buildings, such as steeples, towers, etc., so that small variations, resulting from accidents of the ground, may not have any action on them. Clouds indicate the direction of the upper aerial currents, and show that it differs very often from the direction of the wind on the surface of the earth.

387. That the direction of the wind in the upper regions is often the reverse of what it is in the lower, has been conclusively proved. During an eruption of the volcano of St. Vincent in 1812, the ashes were conveyed in great quantities to the island of Barbadoes, situated to the east. These islands lie in the range of the trade-winds, which blow from the east toward the west; but the ashes, having been launched into the air as high as the region of the upper current, were transported by it in the direction from west to east. At the summit of the Peak of Teneriffe almost all travelers have found west winds, while east winds prevail at the level of the sea. On the 25th of February, 1835, the ashes emitted from the volcano of Coseguina, in the state of Guatemala, obscured the light of the sun for five days; they rose into a high region of the atmosphere, and fell a short time afterward in the streets of Kingston, in Jamaica, which is situated to the N. E. of Guatemala, the winds near the surface all the while blowing toward the S. W.

388. Daily experience teaches us the unequal force of the wind, exhibiting every conceivable variety, from the almost insensible breeze to the hurricane which prostrates the monarch of the forest. The following facts respecting the velocity and force of winds have been ascertained:

Velocity of the wind in miles per hour.	Perpendicular force on one square foot in pounds avoirdupois.	Characteristics.
1	.005	Hardly perceptible.
2	.020	Just perceptible.
5	.123	Gentle, pleasant wind.
10	.492	Brisk gale.
20	1.968	Very brisk.
30	4.429	High wind.
40	7.873	Very high wind.
50	12.300	A storm.
60	17.715	A violent storm.
80	31.490	A hurricane.
100	49.200	A violent hurricane.

389. Winds may be divided into three classes,—Variable, Permanent, and Periodical.

390. VARIABLE WINDS.—Variable winds, as their name indicates, are very irregular as to time, direction, and force, and seldom continue to blow for many days. They prevail in the temperate and frigid zones, those of the torrid zone being, for the most part, either permanent or periodical.

391. We are not fully acquainted with the causes which produce these partial and ever-fluctuating aerial currents, but there is no doubt that they are mainly due to the unequal states of the temperature of land and sea. Although these winds alternately come from every point of the compass, changing frequently from one point to the opposite in a very short space of time, it has been observed that different seasons are characterized by winds from different directions. Franklin long ago observed in North America that in summer the winds come from the south and in winter from the north.

392. From numerous observations made in different parts of Europe, the following laws have been established: In *winter,* the direction of the wind is principally from the south, its force being greatest in January. In *spring,* east winds are common at certain places in March, and at other places in April. In *summer,* especially in July, the winds blow chiefly from the west, and in *autumn* the south winds more frequently blow, particularly in October.

393. Designating the total number of winds that blow in a given time by 1000, the following table shows their relative frequency in the countries named:

Countries.	N.	N. E.	E.	S. E.	S.	S. W.	W.	N. W.
North America	96	116	49	108	123	197	101	210
England	82	111	99	81	111	225	171	120
France	126	140	84	76	117	192	155	110
Germany	84	98	119	87	97	185	198	131
Denmark	65	98	100	129	92	198	161	156
Sweden	102	104	80	110	128	210	159	106
Prussia	99	191	81	130	98	143	166	192

394. There is a certain class of variable winds known to possess peculiar properties, such as the hot winds felt on the northern coast of Africa, in Persia, India, and China, the cold winds of Siberia, the pestilential simoon or samiel of Africa, Arabia, and Mesopotamia, etc. Winds partake of the property of the regions from which they come; thus, in Europe, the west winds which blow from the Atlantic are more moist than the east, which sweep over the continent. On the Atlantic coast of the United States, the winds which come from the northeast are remarkable for their chilliness, and for the disagreeable storms which accompany them. Variable winds may be subdivided into cold and hot winds.

395. The *cold* winds of the north temperate zone are those which blow from the north and northeast. In Europe the northeast winds are cold, deriving their character from the very low temperature which prevails in northern Europe and Asia. In the south of Europe the north winds are of great violence and severity, owing to the contrast between

Questions.—385. Into how many parts is the horizon divided to indicate the direction of the wind? How is the wind designated? What are the eight kinds of winds? Their initials? 386. How is the general direction of the winds near the surface of the earth indicated? Where are they commonly placed, and why? What do clouds indicate? 387. What is said of the reverse direction of upper and under currents? Of the eruption of the volcano of St. Vincent? What have travelers found at the summit of the Peak of Teneriffe? What further illustration of this fact was afforded by the transportation of ashes emitted from the volcano of Coseguina? 388. What is said of the unequal force of the wind? Describe the force, and mention the characteristic, of wind having a velocity of 1 mile per hour, 2 miles per hour, 5 miles per hour.

Questions.—10 miles per hour. 20 miles per hour. 30 miles per hour. 40 miles per hour. 50 miles per hour. 60 miles per hour. 80 miles per hour. 100 miles per hour. 389. In what three classes may winds be divided? 390. What are variable winds? Where do they prevail? 391. To what are they no doubt mainly due? What has been observed respecting them? What did Franklin observe? 392. What is the general direction of the winds of Europe in winter? In spring? In summer? In autumn? 394. What is said of the properties of certain winds? Of the west winds in Europe? Of the northeast winds on the Atlantic coast of the United States? How may variable winds be subdivided? 395. What is said of the cold winds of the north temperate zone? Of the northeast winds in Europe? Of the north winds in the south of Europe?

snow-covered Alps and the elevated temperature of the Mediterranean.

396. The bora, a northeast wind, so called in Istria and Dalmatia, is sometimes so furious as to overturn horses at plow. The mistral and the vent
de bise are winds which prevail in the southeast of France. The mistral
blows from the northwest, descending from the mountains of central France,
and sweeping over the ancient provinces of Provence and Languedoc, where
it is supposed to contribute greatly to the salubrity of the air by dispelling
the exhalations from the marshes and stagnant waters common in that
region of extensive levels. It is very fearful in the Gulf of Lyons; hence
the name of that gulf, not derived, as commonly imagined, from the city
of Lyons, but from the lion-like violence of its tempests. The vent de bise
(black wind) is a cold, piercing current from the Alps and the mountains
of Auvergne, which chiefly follows the course of the Rhone, in the valley
through which it runs, rendering the climate in winter very severe. In
Spain, a north wind, called the gallego, is of a very formidable character.

397. Hot and dry winds are very frequent in countries con-
tiguous to the tropical regions. Large deserts and plains,
covered with little vegetation, engender very warm winds;
these winds, which are of a noxious character, prevail in the
vast deserts of Asia and Africa, where they show themselves
in all their force. Nubia, Arabia, Persia, and other parts of
Asia, are visited by a scorching wind peculiar to the desert. In
Arabia it is called samoun, from the Arabic samma, which
signifies hot and poisonous. It is also named samiel, from
samm, poison. In Egypt it is called chamsin (fifty) because it
blows for fifty days, from the end of April until June, at the
commencement of the inundation of the Nile. In the western
part of the Sahara it is named harmattan.

398. The simoon is announced by the troubled appearance
of the horizon; afterward the sky becomes obscured, and the
sun loses its brilliancy,—paler than the moon, its light no
longer projects a shadow; the green of the trees appears of a
dirty blue, the birds are restless, and the affrighted animals
wander in all directions. The rapid evaporation occurring at
the surface of the human body dries the skin, inflames the
throat, accelerates respiration, and causes a violent thirst. The
water contained in the skins evaporates, and the caravan is a
prey to all the horrors of thirst.

399. This hot wind is deleterious in its mildest forms, occasionally destructive, and many a pilgrim on the way to the shrine of the prophet at Mecca, and
merchant to the marts of Bagdad, have perished by its noxious, suffocating
influences. Bruce suffered from it when ascending the Nile, he and his
company becoming so enervated as to be incapable of pitching their tents,
oppressed as well by an intolerable headache. "The poisonous simoon," he
remarks, "blew as if it came from an oven; our eyes were
dim, our lips cracked, our knees tottering, our throats perfectly dry; and
no relief was found from drinking an immoderate quantity of water."

400. "In June, 1813," says Buckhardt, "in going from Siout to Esné, I
was surprised by the simoon in the plain which separates Furschiont from
Berdys. When the wind arose I was alone, mounted on my dromedary,
and at a distance from every tree and habitation. I endeavored to protect
my face by wrapping it in a handkerchief. Meanwhile the dromedary,
into whose eyes the wind drove the sand, became restless, commenced gal-
loping, and caused me to lose the stirrups. I remained lying on the earth
without moving from the spot, for I could not see to a distance of ten
metres, and I wrapped myself up in my clothes until the wind had abated.
I then went in search after my dromedary, which I found at a very great

distance, lying down near a bush that protected his head against the sand
raised by the wind."

401. Volney gives the following complete account of the simoon and its
effects: "Travelers," he states, "have mentioned these winds under the

STORM IN THE DESERT.

name of poisonous winds; or, more correctly, hot winds of the desert. Such,
in fact, is their quality; and their heat is sometimes so excessive that it is
difficult to form an idea of their violence without having experienced it;
but it may be compared to the heat of a large oven at the moment of drawing
out the bread. When these winds begin to blow, the atmosphere assumes
an alarming aspect. The sky, at other times so clear in this climate, becomes dark and heavy; the sun loses its splendor, and appears of a violet
color. The air is not cloudy, but gray and thick; and is in fact with an
extremely subtile dust, that penetrates everywhere.

402. "This wind, always light and rapid, is not at first extremely hot, but
it increases in heat in proportion as it continues. All animated bodies soon
discover it by the change it produces in them. The lungs, which a too
rarefied air no longer expands, are contracted and become painful. Respiration is short and difficult, the skin parched and dry; and the body con-
sumed by an internal heat. In vain is recourse had to large drafts of water;
nothing can restore perspiration. In vain is coolness sought for; all bodies
in which it is usual to find it deceive the hand that touches them. Marble,
iron, water, notwithstanding the sun no longer appears, are hot. The streets
are deserted, and the dead silence of night reigns everywhere. The inhabitants of towns and villages shut themselves up in their houses—and those
of the deserts in their tents, or in pits they dig in the earth—where they
wait the termination of this destructive heat.

403. "It usually lasts three days, but if it exceeds that time it becomes
insupportable. Woe to the traveler whom this wind surprises remote from
shelter! he must suffer all its dreadful consequences, which sometimes are
mortal. The danger is most imminent when it blows in squalls, for then
the rapidity of the wind increases the heat to such a degree as to cause sudden death. This death is a real suffocation; the lungs, being empty, are
convulsed; the circulation disordered, and the whole mass of blood, driven
by the heat toward the head and breast; whence that hemorrhage at the
nose and mouth which happens after death.

404. "This wind is especially fatal to persons of a plethoric habit, and
those in whom fatigue has destroyed the tone of the muscles and vessels.
The corpse remains a long time warm, swells, turns blue, and is easily separated; all of which are signs of that putrid fermentation which takes place
when the humors become stagnant. These accidents are to be avoided by
stopping the nose and mouth with handkerchiefs; an efficacious method is
also that practiced by the camels, who bury their noses in the sand, and
keep them there till the squall is over.

405. "Another quality of this wind is the extreme aridity, which is such
that water sprinkled upon the floor evaporates in a few minutes. By this
extreme dryness it withers and strips all the plants, and by exhaling too

Questions.—396. What is said of the bora? The mistral? The vent de bise? 397.
Where are hot winds very frequent? By what name is the burning wind called in
Arabia? By what other name is it known? What is it called in Egypt, and why? In
the western part of Sahara? 398. How is the simoon announced? How does it affect
the human body? 399. What further is remarked respecting this hot wind? How did
Bruce suffer from it?

Questions.—What does he say of it? 400. Mention some of the particulars related
by Buckhardt. 401. What does Volney say respecting these winds? To what does
he compare their heat? What is the aspect of the atmosphere during the continuance
of the simoon? 402. What changes does it produce on all animate bodies? 403. How
long does it usually last? When is the danger most imminent? 404. To whom is the
wind especially fatal? 405. What other quality does this wind possess? Give particulars.

suddenly the emanations from animal bodies, crisps the skin, closes the pores, and causes that feverish heat which is the invariable effect of suppressed perspiration."

406. The *Harmattan* blows from the northeast, over Senegambia and Guinea, to that part of the coast of Africa lying between Cape Verde, in 15° north latitude, to Cape Lopez, in 1° south latitude, a coast line of upward of two thousand miles. It occurs during December, January, and February, generally three or four times during that season. It comes on at any hour of the day, at any time of the tide, or at any period of the moon, continuing sometimes only a day or two, at other times five or six days, and it has been known to last upward of a fortnight.

407. Extreme dryness is the property of this wind; all vegetation droops and withers, and should the harmattan blow for several days, the leaves of the lemon, orange, and lime-trees become so parched that they may be readily rubbed into dust. Even household furniture cracks, and in many instances flies to pieces. Though this wind is so pernicious in its effects upon vegetable life, yet it is conducive to the health of the human species, by removing dampness from the atmosphere, and counteracting its effects after a long rainy season.

408. The *Sirocco* is a hot southeast wind, prevailing in the Mediterranean, in Italy, and Sicily, but felt most violently in the country around Naples, and at Palermo. It sometimes commences about the time of the summer solstice, but blows occasionally with great force in the month of July. Though usually attributed to the Sahara, it is supposed by some to arise on the arid rocks of Sicily; and hence is far more violent on the north than on the south coast of the island, about Palermo, and also in the neighborhood of Naples.

409. It is thus described by a traveler during his stay at Palermo :

"On Sunday, July 8th, we had the long-expected sirocco wind, which, although our expectations had been raised pretty high, yet I own greatly exceeded them. Friday and Saturday were uncommonly cool, the mercury never being higher than 72½° ; and although the sirocco is said to have set in early on Sunday morning, the air in our apartments, which are very large, with high ceilings, was not in the least affected by it at eight o'clock, when I rose. I opened the door without having any suspicion of such a change ; and indeed I never was more astonished in my life. The first blast of it on my face felt like the burning steam from the mouth of an oven. I drew back my head and shut the door, calling out to Fullarton that the whole atmosphere was in a flame. However, we ventured to open another door that leads to a cool platform, where we usually walk ; this was not exposed to the wind ; and here I found the heat much more supportable than I could have expected from the first specimen I had of it at the other door. It felt somewhat like the subterranean sweating-stoves at Naples, but still much hotter. In a few minutes we found every fiber greatly relaxed, and the pores opened to such a degree, that we expected soon to be thrown into a profuse sweat.

410. " I went to examine the thermometer, and found the air in the room as yet so little affected that it stood only at 73°. The preceding night it was at 72½°. I took it out to the open air, when it immediately rose to 110°, and soon after to 112° ; and I am confident that in our old lodgings, or anywhere within the city, it must have risen several degrees higher. The air was thick and heavy, but the barometer was little affected : it had fallen only about a line. The sun did not once appear the whole day, otherwise I am persuaded the heat must have been insupportable ; on that side of our platform which is exposed to the wind, it was with difficulty we could bear

it for a few minutes. Here I exposed a little pomatum, which was melted down as if I had laid it before the fire. I attempted to take a walk in the street to see if any creature was stirring, but I found it too much for me, and was glad to get up stairs again. This extraordinary heat continued till three o'clock in the afternoon, when the wind changed at once almost to the opposite point of the compass. All nature languishes under the influence of this wind ; vegetation droops and withers ; the Italians suffering from it not less than strangers. When any feeble literary production appears, the strongest phrase of disapprobation they can bestow is, ' It was written in the time of the sirocco.' "

411. The deserts of Asia and Africa are the regions in which the hot or scorching winds prevail ; but in Spain the *Salano*, a wind which is supposed to arise on the plains of Andalusia, throws the majority of individuals into a condition of peculiar languor. In India, which is covered with a rich vegetation, and in Chile, in Louisiana, and in the great level plains (*Llanos*) of the Orinoco, there are certain local winds of a very elevated temperature.

LESSON IV.

PERMANENT WINDS.

412. THE *Trade-winds* are those permanent breezes which prevail within the tropics, and which maintain nearly the same direction and rate throughout the year. Their direction is from the northeast in the northern hemisphere, and from the southeast south of the line ; but it is more decidedly from the east as the equator is approached. They extend generally from about 28° to 30° on each side of the equator, but their limits vary considerably as the sun is north or south of the equator ; their external and internal boundaries are also very different in the Atlantic and Indian oceans. It is only over the wide ocean that the trade-winds can blow uninterruptedly. Between them

A CALM AT SEA.

is a zone styled the Region of Calms, in which thick, foggy air prevails, with frequent sudden and copious rains, attended by thunder and lightning.

413. The trade-winds may be thus explained. The regions bordering on the equator are the hottest on the earth. In consequence of rarefaction, the air there ascends and flows over the colder masses on either side toward the poles, from which

Questions.—406. Describe the Harmattan. When does it occur ? What is said of its irregularity, etc. ? 407. What is the property of this wind ? What are its effects on bodies ? On the human species? 408. What is the Sirocco ? When does it blow ? Its supposed origin ?

Questions.—411. What is said of the deserts of Asia and Africa ? Of the Salano ? Where do other very warm winds prevail ? 412. What are the trade-winds ? What is their direction ? Their limits ? What region lies between them ? 413. How may the trade-winds be explained ?

a colder atmosphere moves to supply its place. Thus two currents are created in each hemisphere, an upper and a lower, but flowing in opposite directions. If the earth did not rotate on its axis, the lower current in the northern hemisphere, or the trade-wind, would be from north to south, and in the southern hemisphere from south to north. The earth, however, rotates from west to east, and the atmosphere surrounding it partakes of this rotary motion; hence these winds become northeast and southeast.

414. The movements of the trade-winds, and the laws by which they are governed, are thus explained by a well-known writer:

"From the parallel of about 30° north and south, nearly to the equator, we have two zones of perpetual winds, viz. : the zone of northeast trades on this side, and of southeast on that. They blow perpetually, and are as steady and as constant as the currents of the Mississippi River—always moving in the same direction. As these two currents of air are constantly flowing from the poles toward the equator, we are safe in assuming that the air which they keep in motion must return by some channel to the place near the poles, whence it came in order to supply the trades. If this were not so, these winds would soon exhaust the polar regions of atmosphere, and pile it up about the equator, and then cease to blow for the want of air to make more wind of.

415. "This return current, therefore, must be in the upper regions of the atmosphere, at least until it passes over those parallels between which the trade-winds are always blowing on the surface. The return current must also move in the direction opposite to the direction of that wind which it is intended to supply. These direct and counter-currents are also made to move in a sort of spiral curve, turning to the west as they go from the poles to the equator, and in the opposite direction as they move from the equator toward the poles.

416. "This turning is caused by the rotation of the earth on its axis. The earth, we know, moves from west to east. Now if we imagine a particle of atmosphere at the north pole, where it is at rest, to be put in motion in a straight line toward the equator, we can easily see how this particle of air, coming from the pole, where it did not partake of the diurnal motion of the earth, would, in consequence of its vis inertia, find, as it travels south, the earth slipping under it, as it were, and thus it would appear to be coming from the northeast and going toward the southwest : in other words, it would be a northeast wind.

417. "On the other hand, we can perceive how a like particle of atmosphere that starts from the equator, to take the place of the other at the pole, would, as it travels north, in consequence of its vis inertia, be going toward the east faster than the earth. It would therefore appear to be blowing from the southwest, and going toward the northeast, and exactly in the opposite direction to the other. Writing south for north, the same takes place between the south pole and the equator. Now this is the process which is exactly going on in Nature; and if we take the motions of these two particles as the type of the motion of all, we shall have an illustration of the great currents in the air, the equator being near one of the nodes,⁎ and there being two systems of currents—an upper and an under—between it and each pole.

418. "Let us turn now to our northern particle, and follow it in a round from the north pole to the equator and back again, supposing it, for the present, to turn back toward the pole after reaching the equator. Setting off from the polar regions, this particle of air, for some reason which does not appear to have been satisfactorily explained by philosophers, travels in the upper regions of the atmosphere, until it gets near the parallel of 30°. Here it meets, also in the clouds, the hypothetical particle that is going from the equator to take its place toward the pole.

419. "About this parallel of 30°, then, these two particles meet, press

against each other with the whole amount of their motive power, produce a calm and an accumulation of atmosphere sufficient to balance the pressure from the two winds north and south. From under this bank of calms two surface currents of wind are ejected; one toward the equator, as the northeast trades—the other toward the poles, as the southwest passage winds—supposing that we are now considering what takes place in the northern hemisphere only.

420. "These winds come out at the lower surface of the calm region, and consequently the place of the air borne away in this manner must be supplied, we may infer by downward currents from the superincumbent air of the calm region. Like the case of a vessel of water which has two streams from opposite directions running in at the top and two of equal capacity discharging in opposite directions at the bottom—the motion of the water in the vessel would be downward : so is the motion of air in this calm zone. The barometer, in this calm region, is said by Humboldt and others to stand higher than it does either to the north or to the south of it; and this is another proof as to the banking up here of the atmosphere and pressure from its downward motion.

421. "Following our imaginary particle of air from the north across this calm belt, we now feel it moving on the surface of the earth as the northeast trade-wind, and as such it continues till it arrives near the equator, where it meets a like hypothetical particle, which has blown as the southwest trade-wind." [The writer here proceeds upon the supposition that the air from the polar region descending at the calm belt uniformly passes to its equatorial side, and so enters the trade-wind region. It is proper to observe, however, that among a majority of scientific men the evidence that this is the case is deemed inconclusive; although the probability that such is its frequent course is by no means disproved.] "Here, at this equatorial place of meeting, there is another conflict of winds, and another calm region, for a northeast and southeast wind can not blow at the same time in the same place. The two particles have been put in motion by the same power; they meet with equal force, and therefore, at their place of meeting, are stopped in their course. Here, therefore, there is also a calm belt.

422. "Warmed by the heat of the sun, and pressed on each side by the whole force of the northeast and southeast trades, these two hypothetical particles, taken as the type of the whole, ascend. This operation is the reverse of that which took place at the other meeting near the parallel of 30°. This imaginary particle now returns to the upper region of the atmosphere again, and travels there until it meets, near the calm belt of Cancer, its fellow-particle from the north, where it descends as before, and continues to flow toward the pole as a surface wind from the southwest." [Here, again, the writer proceeds upon the supposition that the descending particle of air crosses the calm belt, and continues its course on the opposite side—a theory regarded as resting upon an uncertain basis.

423. The general circulation of the atmosphere which manifests itself thus uniformly in the torrid regions and with less constancy in the temperate, has been a subject of profound scientific investigation. The laws which govern it are complex and obscure; but it is believed that the daily increasing observations, especially those in the higher latitudes, will throw light upon the subject, and in process of time will unravel its most perplexing mysteries.

424. Nothing excited the wonder of the early navigators so much as the trade-winds which blow regularly within the tropics. The companions of Columbus were terrified when they found themselves driven on by continuous easterly breezes, which seemed to forewarn them that they would never return to their country. Fortunately for the fame of the great navigator, and for the world, he firmly held on his course, and made the discovery of a new continent.

425. The trade-winds serve important uses to navigators, in facilitating the passage of ships round the world. In passing from the Canaries to Cumana, on the north coast of South America, it is scarcely ever necessary

---

Questions.—414. Where have we two zones of perpetual winds? What are we safe in assuming? Why? 415. Where must this return current be? Which way do the direct and counter-currents turn? 416. How is this turning caused? Explain. 417. Explain the direction of the return current. 418. Describe the course of a particle of atmosphere proceeding from the polar regions toward the equator. What does it meet near the parallel of 30°? 419. What results follow? What surface currents are here ejected? 420. From what part of the calm region do these winds come, and what consequently may be inferred?

⁎ Nodes, the point where the ascending and descending currents cross each other.

Questions.—Illustrate the downward motion of the air in this calm zone. What is another proof of the banking up here of the atmosphere and pressure from its downward motion? 421. Where is the particle of air supposed to move after leaving the calm belt? What occurs at the equatorial place of meeting? 422. Why does it here ascend? Describe its track as it returns toward the pole. 423. Investigation of the general circulation of the atmosphere? Laws which govern it, and prospect of a future explanation of the subject? 424. Wonder of the early navigators? The companions of Columbus? 425. Uses of the trade-winds? Illustrate. Route of an outward and return voyage from New York to Canton?

3° and 10° south of the line; and when the wind north of the equator is northeast, that south of it is northwest.

430. The western boundary of the region of the monsoons is the African shore; its eastern limit is supposed to be about the meridian of 150° east longitude; its northern confine is near the parallel of 27° north latitude; its southern extremity has been already stated. The monsoons are much stronger than the trade-winds, and may be called gales; they sometimes blow with such violence that ships are obliged to reef their sails. They are not confined to the ocean, but extend over the whole of Hindoostan to the Himalaya Mountains.

431. Mr. Caunter, a resident of Madras, gives the following interesting account of a storm which occurred there during the shifting of these winds:

"On the 15th of October the flag-staff was struck, as a signal for all vessels to leave the roads, lest they should be overtaken by the monsoon. On that very morning some premonitory symptoms of the approaching war of elements had appeared. As the house we occupied overlooked the beach, we could behold the setting in of the monsoon in all its grand and terrific sublimity. The wind, with a force which nothing could resist, bent the tufted heads of the tall, slim cocoa-nut trees almost to the earth, flinging the light sand into the air in eddying vortices, until the rain had either so increased its gravity or beaten it into a mass, as to prevent the wind from raising it.

432. "The pale lightning streamed from the clouds in broad sheets of flame, which appeared to encircle the heavens as if every element had been converted into fire, and the world was on the eve of a general conflagration; while the peal, which instantly followed, was like the explosion of a gunpowder magazine. The heavens seemed to be one vast reservoir of flame, which was propelled from its voluminous bed by some invisible but omnipotent agency, and threatened to fling its fiery ruin upon everything around. In some parts, however, of the pitchy vapor, by which the skies were by this time completely overspread, the lightning was seen only occasionally to glimmer in faint streaks of light, as if struggling, but unable, to escape from its prison—igniting, but too weak to burst, the impervious bosoms of those capacious magazines in which it was at once engendered and pent up.

433. "So heavy and continuous was the rain, that scarcely anything save those vivid bursts of light, which nothing could arrest or resist, was perceptible through it. The thunder was so painfully loud that it frequently caused the ear to throb; it seemed as if mines were momentarily springing in the heavens, and I could almost fancy that one of the sublimest fictions of heathen fable was realized at this moment before me, and that I was hearing an assault of the Titans. The surf was raised by the wind and scattered in thin billows of foam over the esplanade, which was completely powdered with the white, feathery spray. It extended several hundred yards from the beach; fish, upward of three inches long, were found upon the flat roofs of houses in the town during the prevalence of the monsoon—either blown from the sea by the violence of the gales, or taken up in the water-spouts, which are very prevalent in this tempestuous season.

434 "When these burst, whatever they contain is frequently borne by the sweeping blast to a considerable distance over-land; and deposited in the most uncongenial situations; so that now, during the violence of these tropical storms, fish are found alive on the tops of houses; nor is this any longer a matter of surprise to the established resident in India, who sees every year a repetition of this singular phenomenon. During the extreme violence of the storm, the heat was occasionally almost beyond endurance, particularly after the first day or two, when the wind would at intervals entirely subside, so that not a breath of air could be felt, and the punka afforded but a partial relief to that distressing sensation which is caused by the oppressive stillness of the air so well known in India."

435. The monsoons are of great assistance to commerce; by

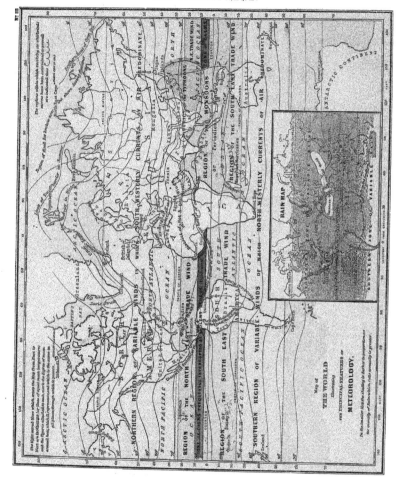

Map of
THE WORLD
Illustrating
THE PRINCIPAL FEATURES OF
METEOROLOGY.

them a ship is frequently wafted to a distant port, and aided in returning by a monsoon blowing in an opposite direction to that which carried her out.

436. The *Etesian winds** are periodical winds which blow from the northeast for about six weeks throughout the Mediterranean, particularly the Levant, where they commence toward the middle of July, about nine in the morning, and continue only throughout the day. The immense desert of Sahara, south of the Mediterranean, deprived of water, and composed of sand and flints, becomes very highly heated under the influence of an almost vertical sun, and currents are created from the colder atmosphere of the north. Hence the passage from Europe to Africa in summer is much quicker than the autumn. Periodical currents, called *nortes*, or north-winds, blow from September to March in the Gulf of Mexico. They occur also on the Brazil coast, from N. E. in the spring, and S. E. in autumn.

437. *Land and sea breezes* are supposed to be caused by the unequal heating of the land and sea. On the coasts and islands within the tropics, a breeze from the sea daily occurs about nine o'clock in the morning—at first blowing gently toward the shore, but gradually increasing in force till the middle of the day, when it becomes a brisk gale; after two or three o'clock it begins to subside, and is succeeded at evening by a breeze from the land, which blows freshly from off the coast during the night, and dies away in the morning when the sea-breeze commences.

438. These breezes are particularly strong along the coast of Malabar, where it is said that their influence is felt 60 miles from land. They are also very perceptible in the Mediterranean and in the East and West India Islands. The regular inland breezes experienced in the morning and evening, in some situations, are produced by changes in the density of the atmosphere, affected by the radiating properties of neighboring snow-clad mountains, marshes, or sandy deserts.

439. The *Zone of Calms*, represented on the map, has a breadth of about 6°, but varies with the seasons from 9° to 10°. It is frequently interrupted by violent storms. "When the vessel on its voyage to the south approaches the equator in the midst of the Atlantic Ocean, anxious fear seizes the crew. Sooner or later, according to the time of year, the favoring wind which had brought them thus far, becomes weaker and weaker; at first it ceases for a little while, and at last drops entirely. Around extends the sea, an endless glassy surface. The ship, hitherto speeding onward with a bird-like flight, lies bound on the crystal fluid. The rays of the sun, falling perpendicularly, glow through and through the narrow space in which the men are inclosed. The deck burns through the soles of the shoes. A stifling vapor fills the cabins. A fortnight has the ruler of the sea lain immovable in the same spot. The store of water is exhausted. Glowing thirst glues the parched tongue to the palate. Each man looks upon his companion in suffering with the wild, murderous glance of despair.

440. "The sun sinks below the horizon, the evening sky is illumined by a peculiar coppery redness; and with the advancing night arises a black wall to the eastward; a low, shrill pipe resounds from the distance, from whence a streak of foam advances over the black ocean. The ship sways and rocks upon the irregular waves, but the sail still hangs against the mast, flapping dismally upon the spars. Suddenly the storm bursts over with frightful roar; with a shriek the sails are torn asunder, and fly in ribbons! A loud crack! a second, and the mainmast goes overboard! By a violent effort the crew succeed in cutting through the remaining ropes, and the ship

* *"Ετος*, a year, a season

now flies over the ocean—now borne high upon the backs of the waves—now hurled down into the depths, so that every seam cracks and groans as though it would part asunder. The thunder rolls unceasingly; continuous lightning darts through the agitated atmosphere; the rain falls in streams instead of drops. Ten times the sailors give themselves up for lost, when the quaking bark falls into the trough of the sea, and as many times does it rise over the waves again.

441. "At last the storm lulls; single shocks follow, always at longer intervals; the waves become smoother, and when the consoling sun rises in the east, it illuminates the same dreary picture as on the former day. Mirror-like the endless surface again expands, and in eight days is the store of collected water exhausted; and again the silent specters creep about and turn murderous looks upon each other. A new storm, and a new calm, and so in frightful alternation, until at last the ship is driven into the region of the peaceful trade-wind on the other side of the equator. Hundreds of ships have gone down in storms here; hundreds lost their crews by the most frightful of deaths,—that of thirst; and those who have passed the fearful region of calms, turn in earnest worship to Heaven with thanks for their new-won life." *

LESSON VI.

HURRICANES.

442. THE terms hurricane, whirlwind, water-spot, land-spout, sand-pillar, tornado, white squall, pampero, etc., have been ap-

plied to rotary movements of the atmosphere in different parts of the world.

443. *Hurricanes* are revolving storms which occur chiefly in the West Indies and in the Indian Ocean. Of a similar kind are the *typhoons* in the China Sea. These circulating movements occupy a space from 50 to 500 miles in diameter. They revolve the more rapidly the nearer the center, up to a certain distance, *within which there is a calm.*

444. The center of rotation advances steadily along a definite line upon the globe, with a velocity varying from 2 to 30 or 40 miles per hour. It is a remarkable fact that in the same hemisphere these whirling storms always revolve the same way, but that this direction is opposite in opposite hemispheres. In the northern hemisphere their rotation is *retrograde,* or in a direction opposite to the hands of a watch. In

the southern hemisphere their rotation is direct, conformable to the hands of a watch.

445. There are three well-known hurricane regions—the West Indies, the Indian Ocean, and the China Sea. The general course of the West Indian hurricanes is from the Leeward Islands N. W., passing around the shores of the Gulf of Mexico or across it, then following the Gulf Stream and terminating in the Atlantic, or exhausting their fury in the United States. From October 3, 1780, to August 25, 1837, inclusive, 38 hurricanes occurred in this region in the following months: in June, 1; July, 4; August, 13; September, 10; October, 8; exclusive of two, the months of which have not been recorded. Thus they are very rare in June, and most frequent in August. The only instance in June occurred in 1831, when Trinidad, Tobago, and Grenada were devastated, before supposed to be exempt from hurricanes.

446. The hurricanes of the Indian Ocean come from the N. E., near Sumatra and Java, and travel to the S. W. toward Rodriguez and the Mauritius. They occur chiefly from December to April, the hot season in that hemisphere; are very rare in November and May; and are quite unknown during the other months of the year.

447. In the China Sea, hurricanes, there styled *typhoons,* range from 10° to 30° N. They occur from June to November, after an interval of three or four years.

NOTE.—*The map-questions relative to "Winds," on page 55, should receive attention before proceeding with the next lesson.*

LESSON VII.

MOISTURE.

448. IF we place a vessel of water in the open air on a warm day, the quantity of the fluid will soon be sensibly diminished,

or evaporated, being converted by the heat into invisible vapor, and diffused through the air. In like manner evaporation transpires upon a grand scale from the great collections of water on the earth's surface, the oceans, lakes, and rivers, as well as from moist ground. It is subject to diurnal and annual variations.

449. The quantity of vapor diffused through the air is least in the morning before sunrise. As the temperature rises with the ascent of the sun, evaporation increases. The heat of mid-

Questions.—442. What terms have been applied to the rotary movements of the atmosphere? 443. What are hurricanes? What space do these circulating movements occupy? Where do they revolve the more rapidly? 444. What is the velocity of the center of rotation? What is a remarkable fact? What is the rotation of the storms in the northern hemisphere? In the southern hemisphere? 445. Which are the three well-known hurricane regions?

Questions.—What is the general course of the West Indian hurricanes? When are hurricanes in this region most rare, and when most frequent? 446. Describe the hurricanes of the Indian Ocean. When do they chiefly occur? 447. What is said of the typhoons? When and how often do they occur? 448. What will happen if we place a vessel of water in the open air on a warm day? Where does evaporation transpire on a grand scale? 449. When is the quantity of vapor diffused through the air the least? When does evaporation increase? What does the heat of mid-day cause? In what month is the quantity of vapor at its minimum, and in what at its maximum?

* Schleiden's "Lecture about the Weather."

day causes the vapors to rise into the upper regions, and hence the greatest degree of dryness is generally felt during the daytime, though evaporation is then going on most rapidly. In January the quantity of vapor, like the mean temperature of the air, is at its minimum: it increases from that period, and in July attains its maximum; it then decreases to the end of the year.

450. The moisture in the atmosphere is an element on which the life of plants and animals as much depends as on temperature. The healthfulness of climate is greatly influenced by the dryness or humidity of the atmosphere. As resulting from the action of heat on water, the quantity of vapor diminishes with the temperature from the equator to the poles. It decreases also as we pass from coasts into the interior of continents. This rule is confirmed in the interior of the United States, in the middle of the plains of the Orinoco, in the steppes of Siberia, the deserts of Asia and Africa, and the central parts of Australia. In the temperate zone, in general, the annual evaporation is estimated at between 36 and 37 inches of water. In the torrid zone, at Guadaloupe, it has been found to amount to 97 inches, and at Cumana to 100 inches.

451. The air is only capable of receiving a certain quantity of vapor. Its capacity depends upon its temperature, and is invariable in its extent at the same temperature. According to Professor Leslie, air at the freezing point is capable of holding moisture equal to the 160th part of its own weight; at the temperature of 59°, the 80th part; at that of 86°, the 40th part; at 113°, the 20th part; and at that of 140°, the 10th part.

452. When a volume of air contains as much aqueous vapor as at its particular temperature it is capable of receiving, it is then said to be at the point of saturation, being as humid as can be. If the temperature then rises, it will be capable of receiving more; but if it falls, some of the contained vapor will be rejected, and become visible as mist. Thus the effect of a change of temperature upon a saturated volume of air is analogous to that of the hand relaxing or tightening its grasp on a piece of imbibing sponge.

453. It has been frequently observed that the summits of some mountains are constantly covered with clouds. The formation of such clouds may be thus explained: The winds, laden with vapor from lower and warmer districts, in passing over such ranges are forced up into a colder region, where they are obliged to part with a portion of the vapor, which thus forms a cloudy state of the atmosphere.

454. Mists and fogs are formed when the air is saturated, and generally when the moist soil, or the water of lakes and rivers is warmer than the air, the vapors of which are immediately condensed. In like manner the vapor of the air breathed from our mouths in winter becomes condensed and visible. Mists differ in no respect from clouds except in position, being on the surface of the earth, instead of being suspended at a height in the atmosphere.

455. The thick mists which prevail in the neighborhood of Newfoundland

arise from the warm waters of the Gulf Stream, which flow to that locality, the temperature of which is much higher than that of the saturated air.

MOUNT EREBUS.

456. Soon after sunset, in calm and clear weather, mists are frequently formed over the beds of lakes and rivers, while the adjacent land is free from them. This arises from the land more rapidly losing its heat by radiation than the lake or river. The air over the land necessarily becomes the coldest; and when the situation of the ground is such as to bring the cold air of the land over the warmer water, a fog confined to its expanse ensues.

457. Dew is formed by the gradual condensation of the vapors of the atmosphere. After sunset in summer, when the great diurnal evaporation has filled the air with moisture, and the earth is gradually cooled by radiation under a clear sky, the atmosphere in contact with the surface is chilled, and has its capacity to retain the aqueous vapors lessened, which are gradually and gently deposited in the form of fluid drops, called dew. In some countries dew supplies the place of rain. In Palestine and western Asia in general, where showers are unknown for several months in succession, the dew formed at night moistens the earth and sustains the vegetation, being often so abundant as completely to saturate the tents, baggage, and clothing of travelers exposed to it.

458. As all objects have not the same capacity for radiating heat, some cooling much more rapidly than others, we frequently find certain bodies densely covered with dew, as grass and leaves, while the bare grounds, metals, stones, and wood are comparatively dry. A thermometer laid on a grass-plot on a cloudless night has been upward of 16° lower than another laid at the same time on a gravel walk. Hence there has been a much more copious deposition of dew on the grass than on the walk,—the herbs needing the nourishment thus receiving it in preference to the bare soil—a striking evidence of an all-wise Being operating in the economy of nature.[*]

LESSON VIII.

CLOUDS.

459. CLOUDS are masses of visible vapor like mists, floating in the atmosphere at a distance from the surface of the earth.

They exhibit an endlessly diversified outline, a remarkably varying density, and appear at different elevations. The dense

1, 1. Cirrus. 3. Cumulus. 5. Stratus. 2 2. Cirro-Cumulus. 4. Nimbus.

clouds are usually formed toward noon, when the vapors are raised up by the ascending currents of air, and then condensed by the lower temperature of the upper regions.

460. Though clouds are generally composed of vapor, they may consist of frozen particles. In winter, during severe cold, we can often observe that the vapors which rise are composed of brilliant needles, that glisten in the sun and resemble small flakes of snow. The same thing must take place in the higher regions of the atmosphere. There exist, therefore, snow-clouds and clouds of vapor of water.

461. Notwithstanding the varied aspect of clouds, they may be arranged into three principal classes—the *cirrus*, the *cumulus*, and the *stratus*.

462. *Cirrus—Curlcloud.* Fig. 1.—The *cirrus* (the *cat's-tail* of sailors) is composed of thin filaments, variously disposed, in the form of woolly hair, a crest of feathers, or slender net-work. The cirri appear in the higher regions of the atmosphere, and are the most elevated of the clouds. Viewed from the summits of high mountains, while the traveler looks down upon other forms of clouds, he beholds these still above him, and apparently at as great a distance as when seen from the plains. The appearance of true cirrus, or curlcloud, is supposed to indicate variable weather; when most conspicuous and abundant, to presage high winds and rain; and when the streaming

fibers have pointed in a particular direction for any length of time, the gale may be expected to blow from that quarter.

463. *Cumulus—Stackencloud.* Fig. 3.—This form of cloud (*ball of cotton* of sailors) occurs in the lower regions of the atmosphere, and is easily recognized. It is commonly under the control of the surface winds, and frequently exhibits a very magnificent appearance. It consists of a vast hemispherical or conical heap of vapor rising gradually from an irregular horizontal base and increasing upward. Hence the names *cumulus*, a pile or heap, and *stackencloud*, a number of detached clouds stacked into one large and elevated pile.

464. Cumuli are indications of fine weather. They begin to form soon after sunrise from irregular and scattered specks of cloud; as the morning advances the clouds enlarge; and early in the afternoon, when the temperature of the day is at its maximum, the cumulus attains its greatest magnitude. The cloud decreases as the sun declines, and is usually broken up toward sunset. The cumulus may be called the cloud of day, from the interval between morning and evening generally measuring the term of its existence. Its appearance considerably varies in the detail, and often exhibits a brilliant silvery light, and a copper tinge, when in opposition to the sun, indicating a highly electrical condition of the atmosphere.

465. *Stratus—Fallcloud.* Fig. 5.—This cloud consists of horizontal bands near the surface of the earth. It belongs to the night, forming at sunset and disappearing at sunrise. This class of clouds comprehends all those fogs and creeping mists which sometimes spread like a mantle over the surface of the valleys, plains, lakes, and rivers.

466. To the above primary varieties three transition or composite forms are added. *Cirrocumulus—Sondercloud.* Fig. 2. This name designates the feathery accumulated cloud, familiarly known as fleecy, intermediate between cirrus and cumulus.

NIGHT SCENE, TICONDEROGA.

It consists of small patches, arranged in extensive beds, the component parts being quite distinct or asunder. *Cirrostratus*

Questions.—459. Of what are clouds composed? What two kinds? 461. Into what three principal classes may clouds be arranged? 462. Describe the cirrus or curlcloud. What is said of the height of the cirri? What is the cirrus supposed to indicate? 463. Describe the cumulus or stackencloud. Of what does it consist?

Questions.—464. What do cumuli indicate? Describe their successive changes during the day. 465. Of what does the stratus or fallcloud consist? When does it prevail? What does this class of clouds comprehend? 466. Describe the cirrocumulus or sondercloud. What other composite forms are mentioned?

—*Wanecloud* and *Cumulostratus—Twaincloud* (not represented in the cut), are combinations,—the former of the cirrus and stratus, and the latter of the cumulus and stratus.

467. *Nimbus—Raincloud.* Fig. 4.—Any of the preceding modifications may pass over into the actual rainy clouds, first exhibiting a great increase of density, and a bluish-black tone of color, then putting on a lighter shade, or gray obscurity, and becoming fringed at the edges.

468. Clouds are generally higher within the tropics than in the temperate zones; and in the temperate zones they are commonly higher in summer than in winter. The cirri are the highest clouds. They are supposed to range from three to five miles above the level of the earth. Kaemtz states that during a stay of eleven weeks within sight of the Finsteraarhorn, upward of 14,000 feet high, he never observed any cirri below the summit of the mountain. It is highly probable that they consist of flakes of snow.

LESSON IX.

RAIN.

469. RAIN is water, which, originally taken up into the atmosphere in the state of vapor, is returned to the earth in the form of liquid drops. It is produced by the continued condensation of vapor. Rain may begin to fall, and yet not reach the ground, being resolved into invisible vapor on arriving at strata of air removed from the point of saturation. For the same reason, rain-drops may become smaller in their descent, a portion being evaporated, and less rain arrive at the earth's surface than at a certain height above it. Usually the drops increase in their fall, bringing with them the low temperature of the upper regions.

470. The following are examples of very extraordinary showers:
1822.—Oct. 25, at Genoa, 30 inches of rain fell in 24 hours.
1827.—May 27, at Geneva, there fell 6 inches of rain in 3 hours.
1841.—June 4, at Cuiseaux, a small town in the valley of the Saone, there fell about 11 inches in 68 hours.
At Cayenne, Admiral Roussin collected 10 inches in 10 hours.

471. Rain is very unequally distributed over the regions of the globe. 1. The average annual quantity of rain is the greatest in tropical climates, and diminishes as we recede from the equator to the poles. 2. It decreases as we pass from maritime to inland countries, because the land supplies a less quantity of vapor than the sea. 3. More rain falls in mountainous than in low level districts, lofty heights arresting the clouds, and promoting the condensation of vapor by their cold summits. 4. The result of experiment shows that a greater amount of rain falls while the sun is below than when above the horizon.

472. The diminution of the average annual quantity of rain from the equator to the poles, appears from the following table:.
San Luis de Maranhao............lat. 3° S. is equal to 276 inches.
Parimaribo, Guiana.............. " 6° N. " 229 "

Sierre Leone, Guinea............	lat. 9° N.	is equal to	189 inches.
Island of Grenada, West Indies ..	" 12°	"	108 "
Havana, Cuba..................	" 23°	"	90 "
Calcutta......................	" 22°	"	81 "
Macao, China..................	" 22°	"	68 "
Charleston, South Carolina.......	" 33°	"	47 "
Rome	" 42°	"	39 "
Edinburg.....................	" 55°	"	24 "
St. Petersburg................	" 60°	"	17 "
Uleaborg, Finland.............	" 65°	"	13 "

Annual quantity of rain within the tropics of the New World..115 inches.
Annual quantity within the tropics of the Old World......... 76 "
Average for the tropics generally........................... 95½ "
Annual quantity of rain in the temperate zones of the New
World (United States)................................. 37 "
In the temperate zones of the Old World (Europe)........... 31½ "
Temperate zones generally................................ 34½ "

473. Although the amount of rain is greater within the tropics than in the temperate zones, yet the number of rainy days is less, because two seasons divide the year—wet and dry; and during the dry season entire months frequently pass away without a drop falling or a cloud being seen. In the temperate zones, also, in passing from the tropics to the polar circles, the number of rainy days increases, although the intensity of rain diminishes.

Annual number of rainy days in North of Syria.................			54
"	"	"	Straits of Gibraltar............ 68
"	"	"	Plains of Lombardy............ 90
"	"	"	Buda, Hungary................112
"	"	"	Plains of Germany............141
"	"	"	England and West France 152
"	"	"	Poland......................158
"	"	"	St. Petersburg................169
"	"	"	Netherlands..................170
"	"	"	East of Ireland................208

APPROACHING STORM. SAMBRO LIGHT—ENTRANCE TO HALIFAX HARBOR.

474. Countries situated in the vicinity of the sea receive, as we have remarked, a larger amount of rain than those inland. This is exemplified in the interior of the United States, of the Llanos of the Orinoco, of the Siberian steppes, of Australia, and on comparing the amount of precipitation in inland and maritime countries. It declines from an annual fall of between 30 and 35 inches on the shores of Great Britain and France to from 15 to 18 inches as the borders of Asia are approached. Mountains produce several exceptions to this rule. The annual number of rainy days decreases also with the increased distance from the sea.

West coast of France................................152 days.
Interior of France ....147 "
Kasan, Plains of the Volga .... 90 "
Interior of Siberia................................... 60 "

475. The annual fall of rain in mountainous districts, as compared with that of low and level countries, is sometimes very striking. At Keswick—a mountainous district in England—the annual average depth of rain is 67

Questions.—467. Change to the nimbus or rain-cloud? 468. What is the comparative height of clouds in different zones and seasons? Height of the cirri? What does Kaemtz state? 469. What is rain? How produced? 470. Give examples of very extraordinary showers. 471. What is said of the unequal distribution of rain? Where is the average annual quantity the greatest? Where does it decrease? What is observed of mountainous districts? What does the result of experiment show? 472. What is the annual quantity of rain in different places? The annual quantity within the Tropics of the Old World? Of the New World? Average for the Tropics generally? Annual quantity in the Temperate Zones of the New World?

Questions.—In the Temperate Zones of the Old World? In the Temperate Zones generally? 473. Where is the number of rainy days less than in the Temperate Zones, and why? What is observed in passing from the Tropics to the polar circles? State the annual number of rainy days in different places? 474. What is said of countries situated in the vicinity of the sea? How is this exemplified? With what does the annual number of rainy days decrease? Illustrate. 475. What facts illustrate the difference between mountainous districts and low and level countries with respect to the fall of rain? What is remarked of the description of Judea by the sacred writer?

inches, while on the sea-coast it is not half that amount. On the Great St. Bernard it is 63 inches, and at Paris only 21. "The description of Judea by the sacred writer, contrasting it with the flat lands of Egypt, though not intended to be philosophic, is in harmony with the teachings of science respecting the important part performed by mountains in the general economy of the earth : 'For the land whither thou goest in to possess it, is not as the land of Egypt, from whence ye came out ; but the land whither ye go to possess it, is a land of hills and valleys, and drinketh water of the rain of heaven.' By arresting the course of the clouds, and producing a condensation of aqueous vapor when a warm current of air lights upon their cold summits, the elevations contribute to precipitate the moisture of the atmosphere, often amid a terrible display of electric phenomena—a blaze of fiery horrors, and the echo of heart-thrilling sounds."°

476. In some portions of the world rain is entirely unknown, or occurs so seldom as to be quite a phenomenon. The rainless or nearly rainless regions of the New World comprise portions of California and Arizona, of the Mexican table-land, and of Guatemala, also the coast region of Peru and Bolivia, and much of Patagonia east of the Andes. Those of the Old World comprehend an immense territory, stretching from Morocco, through the Sahara, a part of Egypt, Arabia, and Persia, into Beloochistan, with another great zone, commencing north of the Hindoo-Koosh and Himalayas, including the table-land of Tibet, the desert of Gobi, and a portion of Mongolia.

477. The rains of most tropical countries are periodical,— seasons of extreme humidity regularly alternating with those of excessive drought. The length of time of the rainy season differs in different districts, but lasts generally from three to five months. The periodical rains commence in Panama, on the west coast of America, in the early part of March ; in Africa, near the equator, and on the banks of the Orinoco, they begin in April; in the countries watered by the Senegal, and at San Blas, in California, they begin in June. The violence of these tropical showers may be inferred from the large annual amount of rain, and from its fall being limited to a few months, and to a few hours during the day. The drops are enormous, very close together, and fall with such rapidity as to occasion a sensation of pain if they strike against the skin.

478. In both continents the districts which have their periodical rains are subject to an occasional intermission, and become rainless for considerable intervals, the drought inflicting terrible suffering on man and beast. Such a period happened between the years 1827 and 1830 in the state of Buenos Ayres, and is known by the name of the *gran seco*, or the great drought. This interval was very destructive to animals. The loss of cattle in the province of Buenos Ayres alone was estimated at one million head. Cattle in herds of thousands rushed into the Parana, and being exhausted by hunger, they were unable to crawl up the muddy banks, and thus were drowned.

LESSON X.
SNOW AND HAIL.

479. Snow is vapor somewhat condensed, and congealed before it collects in drain-drops. Snow-flakes exhibit forms

Questions.—476. What do the rainless or nearly rainless regions of the New World comprise? What those of the Old World? 477. What is said of the rains of most tropical countries? Length of the rainy seasons? When do they commence in Panama, etc.? Violence of these tropical showers? 478. To what are the rainless districts of both continents subject? Example? Destruction of cattle? 479. What is snow?

° Rev. Thomas Milner.

of exquisite beauty, regularity, and endless variety. These varied shapes are assumed while the body passes from the vapor form to the solid state. The tendency of vapor to crystallize, while in the process of congelation, may be observed in frost as it collects on the window-panes in winter.

480. A microscope applied to a flake of snow which has fallen in a still atmosphere will unfold its wonderful mode of structure. It is only in the polar regions that snow assumes its most beautiful and varied forms. Captain Scoresby has figured ninety-six different varieties, which he discovered during his arctic voyages, and which he distributed into classes of lamellar,* spicular,† and pyramidal crystals, as shown in the

VARIOUS FORMS OF SNOW CRYSTALS.

above representation. It will be seen that the annexed forms are mostly hexagonal‡ stars, and consequently snow-flakes belong to the hexagonal system of crystals. Kaemtz observes that flakes which fall at the same time have generally the same form; but if there is an interval between two consecutive falls of snow, the forms of the second are observed to differ from those of the first, although always alike among themselves.

481. The limits of the fall of snow at the level of the sea, in the northern hemisphere, are about the parallel of 30° in Amer-

Questions.—What do snow-flakes exhibit? Where are these varied forms assumed? 480. In what region does snow assume its most varied forms? How many varieties discovered by Captain Scoresby, and how distributed? What does Kaemtz observe? 481. Limit of the fall of snow in America? In the center of the North Atlantic? In the Old World?

* LAMELLAR, composed of thin plates or scales.   † SPICULAR, resembling a dart.
‡ HEXAGONAL, having six sides and angles.

ica, which cuts the southern part of the United States; 43° in the center of the North Atlantic; and 36° in the Old World, the latitude of Algiers. But for several degrees above these limits its appearance is rare and brief.

482. Snow performs an important part in the general economy of nature. In winter it serves as a mantle to keep the ground warm, and thus protect vegetation from being destroyed by the frost, or by cold biting winds. Accumulated on elevated mountain chains, it affords, by its thawing, a regular supply to rivers and to the interior reservoirs of the earth, while in low latitudes it tempers the heat of warm regions.

483. *Hail* appears to be partly the result of intense cold rapidly produced in the atmosphere; it is supposed to be also somewhat dependent upon electricity, which is almost always powerfully developed during hail-storms. In very high latitudes it is unknown, and it is also rare at the level of the sea within the tropics. The icy particles which fall vary in shape and size. True hail is an opaque mass, and has generally the form of a pear, or of a mushroom; large hailstones are surrounded by a thick coat of ice, and are composed of alternate layers of snow and ice. No one has ever seen hailstones formed entirely of transparent ice.

484. Many instances are well authenticated of hailstones having a circumference of from 6 to 9 inches, and a weight of from 12 to 14 ounces; but much larger masses are recorded. June 15, 1829, the hail beat in the roofs of the houses at Cazorta, in Spain,—some of the hailstones weighing upward of 4 lbs. avoirdupois. In Hungary, May 8, 1832, a block of ice fell, about 39 inches in breadth and length, and 27 inches in depth. Mr. Darwin mentions a fall of hail in the state of Buenos Ayres which killed a large number of wild animals, ostriches, and smaller birds. These enormous

SNOW STORM.

masses are either the fragments of a thick sheet of ice suddenly formed, and broken in the atmosphere in falling, or are due to the union of a great number of hailstones in their descent.

NOTE.—*The map-questions on pages 63 and 64 should receive attention before proceeding with the following lesson.*

Questions.—482. What are some of the uses of snow? 483. Of what does hail appear to be the result? Upon what is it supposed to be also somewhat dependent? Where is it unknown and where rare? Appearance of true hail? 484. Size of some hailstones observed? Examples of destructive effects of hail? What is remarked of these enormous masses? 485. What is said of climate? What does the term climate commonly denote? Taken in its more general sense, what does it signify?

LESSON XI.

CLIMATE.

485. CLIMATE, in its relation to animal and vegetable existence, constitutes one of the most interesting and important subjects belonging to physical geography. The term, as it is commonly understood, denotes the temperature of the air in the various regions of the globe; but taken in its more general sense, it signifies all those states and changes of the atmosphere which sensibly affect our organs,—temperature, humidity, variation of atmospheric pressure, the purity of the atmosphere, or its admixture with more or less deleterious exhalations, and lastly, the degree of habitual transparency of the air and serenity of the sky, which has an important influence on the feelings and the whole mental disposition of man.

486. Climate is determined by a variety of causes, the chief of which are: 1. The latitude of a country; that is, its geographical position with reference to the equator. 2. Elevation of the land above the sea-level. 3. The proximity to, or remoteness of a country from, the sea. 4. The slope of a country, or the aspect it presents to the sun's course. 5. The position and direction of mountain chains. 6. The nature of the soil. 7. The degree of cultivation and improvement at which the country has arrived. 8. The prevalent winds. 9. The annual quantity of rain that falls in a country.

487. (1) *The latitude of a country*, and the consequent direction in which the solar rays fall upon its surface, are the principal causes of the temperature to which it is subject. At the equator, and within the tropics, the greatest heat is experienced; because the sun is always vertical to some place within those limits, and the solar action is more intense in proportion as the rays are perpendicular to the earth. As we recede from the equator, they fall more obliquely; and because fewer of them are spread over a larger space, they are less powerful, or, in other words, less heating. It has been calculated that, out of 10,000 rays falling upon the earth's atmosphere, 8,123 arrive at a given point if they come perpendicularly; 7,024, if the angle of direction is 50°; 2,821, if it is 7 ; and only 5 if the direction is horizontal.

488. The latitude of a place is therefore of the first importance in determining its temperature, since a decrease of heat takes place with an increase of latitude as we travel, at the same level above the sea, from the equator toward the poles. This is true of countries lying between the tropics and the poles, but it is not true of places situated between the tropics and the equator.

489. "If the ecliptic, as shown on a terrestrial globe, be examined, it will be seen that toward the northern and southern limits, for a considerable distance, it neither approaches nor recedes from the equator or the pole, but has a direction due east and west. This ecliptic is, in point of fact, the path or the point of *direct heat and sunlight* over the earth's

[CONTINUED ON PAGE 64.]

Questions.—486. What are the principal causes which determine climate? 487. What principally determine the temperature of a country? Why is the greatest heat experienced within the tropics? What happens as we recede from the equator? 488. Why is the latitude of a place of the first importance in determining its temperature? Is this true of countries lying between the tropics? 489. Explain why a greater degree of heat prevails at the tropics than at the equator.

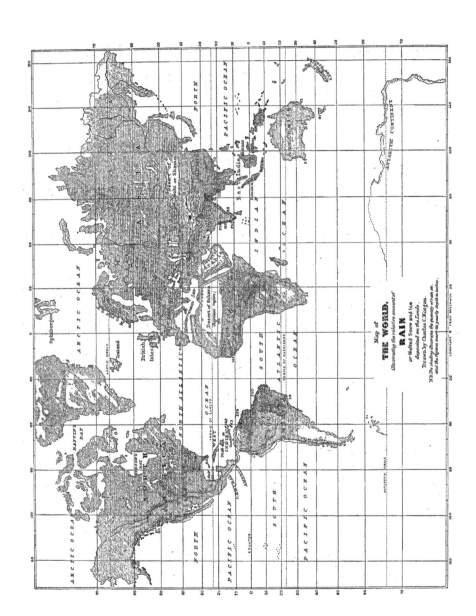

Map of
THE WORLD,
illustrating the relative amount of
RAIN
or Melted Snow and Ice
deposited on the Lands.

Drawn by Charles C. Morgan.

N.B. The shading illustrates the quantity of rain etc.
and the figures mark its yearly depth in inches.

QUESTIONS ON THE ...

DISTRIBUTION OF ... AND ITS INFLUENCE

surface. Thus, then, it appears that when this point has reached its nearest approach to either pole, it does not at once turn back toward the other pole, but remains at that nearest distance for a considerable time; or, as it were, lingers there. It has been calculated, that if the space between the tropics be imagined to be divided into three equal bands of the earth, the point of direct sunlight would be found to linger in each of the two outer bands $8\frac{1}{2}$ times as long as in the middle band.

490. "This lingering of the point of direct heat and sunlight at its nearest approach to the poles, is a necessary result of that simple and admirable provision by which the earth is made to revolve round the sun, rotating at the same time round an axis which has an inclined position, and which preserves its parallelism. The object of it is obviously to minister to the polar regions, in their due proportion, light and heat."○

491. In the northern hemisphere, the countries where the greatest heat is experienced—the banks of the Senegal, the Tehama of Arabia, and Mekran in Beloochistan—coincide with the Tropic of Cancer; and it has been found that the snow-line of the Andes, in 17° south latitude, is higher than at the equator—an evidence of a higher temperature.

492. (2) The temperature of countries is largely affected by the extent of *their elevation above the level of the sea.* As we ascend in the atmosphere the cold increases,—an effect due to the rarefaction of the air, and to the circumstance of being farther from the heat reflected from the surface of the earth. We may travel several hundred miles from the equator toward the poles, along the level surface of the earth, before we become sensible of a diminished temperature; but the moment we begin to increase our elevation, a rapid change of temperature is experienced, until we arrive at a point where constant frost prevails.

POTOSI IN THE ANDES (ELEVATION 13,350 FEET).†

Questions.—490. What is said of the lingering of the point of direct light and heat at the tropics? What is obviously its object? 491. What is said of the countries in the northern hemisphere where the greatest heat is experienced? 492. By what else is the temperature of countries largely affected? What occurs as we ascend in the atmosphere, and why? State the difference, as affecting climate, between traveling on the surface of the earth, from the equator toward the poles, and increasing our elevation.

* Professor Moseley's "Astro-Theology."

† See paragraphs 69 and 70.

493. The ratio of the diminution of temperature usually given, is 1° for 300 feet of height; 2° for 595 feet; 3° for 872 feet; 4° for 1,124 feet; 5° for 1,347 feet; and 6° for 1,539 feet. In the temperate zone generally, if one site is a thousand yards higher than another adjoining, it will have a climate 12° colder; and the higher the latitude the lower the snow-line becomes, till it meets the surface of the earth in the frigid zone. The following diagram represents the line

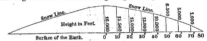

of perpetual snow forming the arc of an ellipsoid passing over the equator, from pole to pole. Making allowance for the fact that the snow-line of the Andes is higher at some distance from the equator, the diagram should not exhibit a continuous curvature, but be corrected as follows:

494. From this effect of elevation upon temperature, it is obvious that the mountainous regions of the torrid zone have great varieties of climate. The hot regions, *tierras calientes*, of Mexico include the country along the eastern and western shores under the elevation of two thousand feet, where the mean temperature is about 77°, and sugar, indigo, cotton, and bananas flourish luxuriantly. Above these are the temperate regions, *tierras templadas*, which lie along the slopes of the mountains at an elevation of from 2,000 to 5,000 feet. Here the yellow fever, the scourge of the low grounds, is unknown ; and the mean heat of the year is from 68° to 70°. The traveler enjoys a genial air, and encounters the oaks, cypresses, pines, tree-ferns, and the cultivated grains of the United States. Still higher are the cold regions, *tierras frias*, which include the table-lands and the mountains above 5,000 feet. On the borders of this zone the climate is still pleasant, but beyond the elevation of 8,000 feet it becomes severe, and gradually assumes the character of polar latitudes.

495. Switzerland, from similar causes, exhibits a variety of climates. In the narrow and deep valley of the Valais may be found great extremes of climate,—the cold of Iceland and the heat of a Sicilian summer. Vines grow to the height of 2,380 feet above the level of the sea ; trees to 6,700 feet ; shrubs to 8,500 ; a few plants to 10,800, beyond which are a few lichens ;. and the vegetation ceases entirely at the height of 11,000 feet, amid Arctic cold.

496. The following is the extreme line of elevation above the level of the sea observed here of individual classes of vegetation :

	Feet.		Feet.		Feet.
Maize	2,772	The cherry	4,270	The silver birch	5,500
The oak	3,518	Potato	4,450	The larch	6,000
The walnut	3,620	The nut	4,590	The fir le sapin	6,300
The yew	3,740	The beech	4,800	Pinus cembra	6,600
Barley	4,180	Mountain maple	5,100	Rhododendron	7,400

497. Etna is divided into three districts, each of which has a climate very different from either of the others. They are called *La Regione Culta*, or the fertile region ; *La Regione Sylvosa*, the woody region ; and *La Regione Deserta*, the barren region. The productions of these districts are as diverse from each other as are those of the three zones of the earth.

Questions.—493. What is the ratio of the diminution of temperature for different heights ? What is remarked of the effect of elevation in the temperate zone ? 494. What is obvious from this effect of elevation upon temperature ? Describe the hot regions, *tierras calientes*, of Mexico. The temperate regions, *tierras templadas*. The cold regions, *tierras frias*. 495. What may be found in the valley of the Valais, Switzerland ? What are the heights to which different kinds of vegetation grow there ? 496. What is the extreme line above the sea-level reached by certain classes of plants ? 497. How is Etna divided ? What are the districts called, and what is said of their productions ?

498. The Island of Teneriffe, with its celebrated Peak, presents another illustration of the effect of elevation on climate. The Peak, which rises to the height of 12,176 feet, has several zones of vegetation, arranged one above another, extending through a perpendicular elevation of 11,190 feet, to which vegetation ascends in that tropical latitude. The region of *Vines* rises from the level of the sea to a height varying from 1,200 to 2,000 feet. In this zone the date-tree flourishes, the sugar-cane, the plantain, the Indian fig, the olive, wheat, and the fruit-trees of Europe. The region of *Laurels* includes the woody part of Teneriffe, in which hardy species of trees, as the oak, the laurel, etc., abound. Next comes the region of *Pines*, filled with trees resembling the Scotch fir. Higher still may be found species of grasses which reach nearly to the line of perpetual snow.

499. The delightful climate of the valleys of Cashmere, and of portions of Hindoostan lying on the declivities of the Himalaya Mountains, is owing to the elevations of those regions above the level of the sea. In these districts, as well as on the table-land between the eastern and western Ghauts, the temperature throughout the year is that of perpetual spring. The inhabitants of Quito, at an elevation of 9,540 feet, experience a genial and almost invariable climate, in which vegetation never ceases, while on the one hand, they behold the mountain ridges covered with perpetual snow, and at the distance of a few leagues an intense and sickly degree of heat oppresses the plains and lower levels.

LESSON XII.

CLIMATE—(CONTINUED).

500. (3) THE *nearness to, or remoteness of a country from, the sea*, is an important element in determining its climate. The ocean preserves a much more uniform temperature than the land, far lower than its extreme of heat, and higher than its extreme cold. The winds that sweep over it have this character to some extent impressed upon them, and communicate it to the countries over which they range. Hence islands and maritime districts have milder climates than inland regions under the same parallel of latitude, the currents from the ocean tempering their summer heat, and moderating their winter cold.

501. London, though situated in a higher latitude, enjoys a milder winter and cooler summer than Paris. The winters and summers of Ireland are much more temperate than those of any other country under the same parallel. A gradual increase of cold is experienced in proceeding from Western Europe in an easterly direction. At Dublin the mean annual temperature is 49.05, while that of Warsaw, in the same latitude, is 44.15 ; thus it is nearly 5° colder at Warsaw than at Dublin. Again, it is 6½° colder at Moscow than at Copenhagen, though they are equidistant from the equator.

502. America has a colder climate than corresponding latitudes in the Eastern Hemisphere. With reference to this, Humboldt advances the following statement: "The comparative narrowness of this continent,—its elongation toward the icy poles,—the ocean, whose unbroken surface is swept by the trade-winds,—the currents of extremely cold water which flow from the Straits of Magellan to Peru,—the numerous chains of mountains abounding in the sources of rivers, and whose summits, covered with snow, rise far above the region of the clouds,—the great number of immense

Questions.—498. What is said of the zones of vegetation on the Peak of Teneriffe ? Describe the region of Vines. Of Laurels. Of Pines. 499. Climate of Cashmere and portions of Hindoostan ? Quito ? 500. What is the third cause which affects climate ? What is said of the ocean ? Climate of islands and maritime districts ? 501. Comparative climate of London and Paris ? Ireland ? What is experienced in proceeding from Western Europe in an easterly direction ? How much colder is it at Warsaw than at Dublin ? At Moscow than at Copenhagen ? 502. How does the climate of America compare with that of the Eastern Hemisphere ? Give the substance of Humboldt's statement.

rivers that, after innumerable courses, always tend to the most distant shores,—deserts, but not of sand, and consequently less susceptible of being impregnated with heat,—impenetrable forests, that spread over the plains of the equator, abounding in rivers; and which, in those parts of the country that are the farthest distant from mountains and from the ocean, give rise to enormous masses of water, which are either attracted by them, or are formed during these acts of vegetation; all these causes produce, in the lower parts of America, a climate which, from its coolness and humidity, is singularly contrasted with that of Africa. To these causes alone must we ascribe that abundant vegetation, so vigorous and so rich in juices and that thick and umbrageous foliage, which constitute the characteristic features of the new continent."

503. (4) *The slope of a country, or the aspect it presents to tue sun's course,* has an important influence on its climate. The angle at which the sun's rays strike the ground, and consequently the power of those rays in heating it, varies with the exposure of the soil relatively to that luminary. The irregular surface of the earth,—sunk into deep valleys in some parts, and raised into table-lands and mountains in others, with slopes at all possible angles with the general level,—presents every variety so far as the greater or less obliquity of the sun's rays is concerned.

504. The effect of aspect is, of course, most strikingly seen in regions covered with high mountains. "Under a vertical sun, the sides of the Andes receive the sun's rays as obliquely as they fall in our latitudes upon the earth's level surface,—nay, as obliquely, perhaps, as they fall in summer upon the level surface of the snows of Spitsbergen; while the Alps encounter on parts of their southern slopes as direct a heat as that which burns up the desert of Sahara; and on their northern they are as much hidden from the sun's influence as are the level snows of Lapland.

505. "In the Alps of the Valais, on the one side you may see the vine in luxuriant growth, when the other is thick ribbed with ice. Thus, too, the terraces and sloping plains which descend from the vast table-land of central Asia, where, inclining from its northern limit, they pass into the steppes of Siberia, present, under the latitude of Edinburg, a cold intense enough to freeze mercury; while upon the southern terraces of the opposite Himalaya slope, flourish, at different elevations, the pineapple, the mango, the gigantic cotton-tree, and the saul. This tropical vegetation ascends there to an altitude of four or five thousand feet; mingling itself, and by degrees giving way to the plants of a temperate region,—elms, willows, roses, and violets, destined in their turn, at a yet higher region, to yield to Alpine forms of vegetable life." ʘ·

506. (5) *The position and direction of mountain chains.*— Mountains affect climate in more ways than one. They condense the vapors of the atmosphere, and thus give rise to those violent rains which are so often experienced in the neighborhood of lofty ranges. At Bergen, in Norway, there fall annually 88½ inches of water, which is more than at any other city in Europe; this is because the clouds from the Atlantic are driven forward by the southwest winds into the fiords, where they are arrested by the mountains, and accumulated, and the water (as it were) mechanically squeezed out of them.

507. Mountains also afford shelter from the winds, while the absence of them often expose regions to the chilling blasts of the north or the burning winds of the south. One reason why the central and southern parts of European Russia are exposed to a greater degree of cold than their latitude would lead us to expect, is the absence of any chain of mountains to

protect them from the full influence of the winds blowing from the White Sea and Ural Mountains. The inhospitable climate

STEPPES OF SOUTHERN RUSSIA.

of Siberia arises from its descent toward the north, exposing it to the winds of the Frozen Ocean; while, at the same time, vast mountain chains that cross central Asia intercept the southern winds, whose access would tend to mitigate the rigor of the atmosphere.

508. (6) Another cause which affects climate is *the nature of the soil.* One soil acquires heat, keeps its acquired heat much longer, or radiates it more readily than another. All the varieties of soil,—light and open, vegetable molds, gravelly and rocky tracts, stiff, wet clays, and sandy plains,—have, it can not be questioned, their different powers of radiation and absorption; and whether a district be clay or sand, bare or covered with vegetation, for a like cause, greatly affects its temperature. The differences of surface so observable in various kinds of foliage,—their darker or lighter colors, their more or less glossy leaves,—are all circumstances which affect the radiation of their heat with an infinite variation.

509. Thus clayey or marshy grounds lower the temperature, and, especially in hot and humid climates, affect the atmosphere in a manner pernicious to health. On the other hand, those which are light, stony, or calcareous, tend to make the atmosphere salubrious. The great cold and the unwholesome air that prevail in the Russian governments of Astrakan and Orenburg, lying to the north of the Caspian Sea, are attributed partly to the saline nature of the soil; and it is well known that the arid tracts of land in Africa and Arabia conduce not a little to the excess of heat under which those countries labor.

510. (7) *The degree of cultivation and improvement at which a country has arrived.*—The clearing of forests, the draining of swamps and marshes, the cultivation of the soil, etc., are among the operations of man by which the climate of a country is greatly modified and improved. The clearing of a country from trees has the effect of raising the mean annual temperature, but at the same time greater extremes of heat and cold are introduced. Open grounds are always frozen deeper than woodlands, but the latter retain the snow and ice of winter to a much later period in the spring than the former. In the cleared portions of North America, the winters are known to be much less severe now than formerly, when woods spread over a greater part of the country. There is little doubt

---

Questions.—How do mountains cause rain? Annual fall of rain at Bergen? How accounted for? 507. What is another way in which mountains influence climate? How illustrated in the case of the climate of southern Russia? From what does the inhospitable climate of Siberia arise? 535. What is the sixth cause which affects climate? Different varieties of soil? 509. Effect of different kinds of ground on temperature? Illustrate. 510. Seventh cause which affects climate? Effect of clearing a country from trees?

* Professor Mosel·y's "Astro-Theology."

that many parts of Europe enjoy a milder climate now than they did in the time of the Romans, or even at periods much more recent.

511. (8) *The prevalent winds* of a country constitute another cause which affects its climate. This is obvious enough from what has been said in the lessons on Winds, from which we learn that the character of a wind depends upon the region whence it comes. In the United States the winds from the north are usually noted for their coolness, a property they derive in the frozen regions of Hudson and Baffin bays, while those from the south, coming from the Gulf of Mexico, impart a mildness throughout the whole country. The comparatively mild climate of the British Isles is owing to the prevalence of westerly winds which are warmed by sweeping over the region of the Gulf Stream. In Venezuela, the temperature, which is from 87° to 90° in March, rises to 104° or 105° whenever the wind blows from the parched surface of the llanos or great plains.

512. (9) *The annual quantity of rain that falls in a country* considerably affects its climate by imparting a greater or less degree of humidity or dampness to the atmosphere. In general, more rain falls in islands and on sea-coasts than in inland districts—among mountains than in level regions, and within the tropics than in the other zones: the great heat which prevails in the equatorial regions causes the amount of evaporation to be much greater than in higher latitudes, and hence the atmosphere becomes loaded with a greater quantity of moisture.

LESSON XIII.

CLIMATE—(CONTINUED).

513. In almost all northern latitudes, January and February are the coldest months of the year, and July and August the warmest. The greatest cold during the day is usually about an hour before sunrise. The greatest heat in latitudes between 35° and 60° is from two to three o'clock P.M., and from one to two o'clock, between the equator and 35°

514. The mean temperature of different months, in various places, takes a very wide range above and below the mean annual temperature. An "excessive" climate is where the contrast between the winter and summer temperature is very great. Such climates are chiefly found in north and eastern Europe, in Asia and America.

515. The climates of the Atlantic region of the United States, and of the northern part of China, are among the most "excessive," the winters and summers strongly contrasting in their temperature. New York has the summer of Rome, and the winter of Copenhagen. At Quebec, grapes sometimes ripen in the open air, whereas the winter is that of St. Petersburg, during which the snow lies from three to five feet deep for several months. At Pekin, in China, the scorching heats

of summer are greater than at Cairo, and the winters as rigorous as at Upsal.

516. The following table gives the difference in the mean temperature of the coldest and warmest months at different places:

Places.	Latitude.	Coldest Month.	Warmest Month.	Difference.	Observations.
Cumana...........	10° 27′	80.1°	84.4°	4.8°	Uninterrupted trade-winds.
Pondicherry......	11 55	76 1	91.4	15.3	Monsoons. Radiation of sands.
Manilla..........	14 36	68.0	86.9	18.9	Monsoons
Vera Cruz........	19 11	70.0	81.7	11.7	North winds in winter.
Cape Français....	19 46	77.0	86 0	09.0	Uninterrupted trade-winds.
Havana.	23 10	70.0	83.8	13.8	North winds in winter.
Funchal..........	32 37	64.0	75.6	11.6	Insular climate.
Natchez..........	31 28	46.9	78.8	31.9	Interior region.
Cincinnati.......	39 06	29.6	74.4	44.8	do. do.
Pekin	39 54	24.8	84.2	59.4	Region of eastern Asia.
Philadelphia.....	39 56	29.8	77.0	47.2	Eastern America coast.
New York........	40 40	25.3	80.8	55.5	do. do. do.
Rome............	41 53	49.1	77.0	34.9	Southern Europe.
Milan...........	45 28	35.8	55.2	24.4	Interior land.
Buda............	47 29	27.6	71.6	43.9	do. do.
Paris...........	48 50	85.1	69.8	34.7	Nearer the western coast.
Nantes..........	47 13	39.0	70.0	31.0	do. do. do.
Upsal...........	60 07	24.2	61.8	87.6	North Europe.
Quebec	46 04	14.0	73.4	59.4	Eastern America coast.
Dublin..........	53 21	87.6	60.3	22.7	Insular climate.
Edinburg........	55 58	38.8	59.4	21.1	do. do.
Warsaw..........	52 14	27.1	70.9	43.2	Interior land.
St. Petersburg ...	59 56	08.6	65.7	57.1	Northern Europe.
North Cape......	71 00	23.1	46.6	94.5	Climate of coasts and islands.

517. *Isothermal* lines (signifying lines of equal heat) are those drawn over places which have the same, or nearly the same, annual temperature. These lines do not coincide with the parallels for the reason that there are other causes besides solar action which influence climate, as explained in the last two lessons. The meteorological map shows, by the use of isothermal lines, those regions which have the same, or nearly the same, temperature. The figures attached to each curve indicate the mean annual temperature of the region traversed by it.

518. In the northern hemisphere, the curve line on the map, indicating the mean annual temperature of 80°, crosses Central America about the

SCENERY ON THE BORDER OF HONDURAS.

Gulf of Honduras, passes south of St. Domingo, descending across the Atlantic to the west coast of Africa, which it cuts near Sierra Leone. It rises on that continent, and after crossing the Red Sea and Arabia, it returns toward the equator, traversing the southern part of Hindoostan, and the northern part of the peninsula of Malacca.

519. The Warmth Equator shows the district of the greatest heat, which have temperatures varying from 81° to 88°. It does not coincide with the geographical, but lies mostly to the north of that line, attaining its greatest distance in eastern Africa and in the Indian Ocean.

520. The isothermal line of 70° crosses northern Mexico and the peninsula of Florida, its course in America being nearly coincident with the parallel of 28°. It reaches the coast of Africa above the Canary Islands, ascends toward Tunis about latitude 34°, runs through the Mediterranean south of Candia, enters Syria north of Beyroot, traverses Asia south of the Hindoo-Koosh and Himalaya mountains, and descends in China to the tropic. This line marks the north boundary of the warm zone, and is generally the southern limit of the fall of snow at the sea-level.

521. The isothermal line of 50°, central to the temperate zone, passes the mouth of the Columbia River, on the west coast of America, latitude 46°, descends in the interior, and reaches the shores of the Atlantic near New York, latitude 41°. It then rises abruptly to latitude 56° in the north Atlantic, descends by Dublin, London, and through midland Europe to the mouth of the Danube, latitude 45°, cuts the north of the Black and Caspian seas, falls in the interior of Asia, rising on the eastern coast.

522. The isothermal line of 30°, indicating an average temperature below the freezing-point, leaves the west coast of America in latitude 61°, rises and declines in the interior, falling to latitude 53° on the shores of Labrador. It then ascends abruptly to latitude 74° in the Arctic Ocean, and passing around the North Cape of Europe, as abruptly descends to latitude 59° in the interior of Asia. This line marks the southern limit of permanently frozen ground, which, in Asia, occurs in a latitude as low as that of London.

523. The isothermal lines indicating a lower degree of temperature can only be traced with certainty through portions of their course. That of 20° appears to descend from the mouth of the Mackenzie River, latitude 68° to latitude 59° on the coast of Labrador. From thence it rises to latitude 77°, cutting the south of Spitzbergen, and falls in Asia to latitude 56°. The isothermal line of 10° passes to the north of Hudson Bay in America, and of Yakutsk in Asia.

524. The western part of Europe has much greater warmth than corresponding latitudes in eastern America, as is indicated by the convexity of the isothermal lines about the meridian of Greenwich. This is due to a variety of causes. 1. The prevailing winds being from the southwest, come from the equatorial districts, and partially bear the heat of the tropics to the shores of Europe. 2. The warm waters of the Gulf Stream, which sweep across the Atlantic to the neighborhood of Europe, elevate the temperature of the coasts that are washed by them. 3. The Gulf Stream, it is true, runs along the east coast of America from Florida to Cape Hatteras, but the current is there very narrow, and the prevailing winds carry the warm superincumbent air away from the shore. On the other hand, a sea current comes down from the Arctic Ocean to the coasts of Labrador and Newfoundland, and is traceable some distance southward between the Gulf Stream and the coast, which contributes to depress the temperature by its cold waters, but chiefly by the floating icebergs that descend with it as far as Newfoundland, and in some cases even below this latitude.

525. With reference to climate and productions, the globe may be divided into the following regions,—the hot, warm, temperate, cold, and frigid.

526. The Hot Regions occupy a zone extending on each side of the equator to a few degrees beyond the northern and southern tropic. Here abound the finest spices, the sugar-cane, the palm and banana tribes. It includes the islands and mainland of southern Asia, the middle and northern countries of Africa, and the central parts of America.

527. The Warm Regions extend from the northern limit of the sugar-cane to the northern boundary of the olive and fig; and include the southern districts of Europe, part of mid-land and southwestern Asia, and portions of the southern United States. The frosts here are not severe in the plains; snow is rare; and the rivers are seldom frozen over. The winters are more distinguished for dampness than cold, resembling the spring of the temperate regions. Vegetables on the south of this zone grow during seven or eight months of the year, and the trees are stripped of their foliage more than two months.

528. The Temperate Regions reach from the northern region of the olive and fig to that of the grapevine, and have a mean temperature varying from 50° on the northern border to 50° on the southern. The transition from winter to summer is here gradual, and the four seasons are distinctly marked. Grain, vegetables, and many fine fruits are produced in abundance, with excellent pasturage. The northern part of the United States is included in this region, with France, Germany, southern Russia, Holland, Belgium, England, and Ireland, in Europe.

529. The Cold Regions include the country between the northern limits of the grapevine and the oak, or midland

INTERIOR OF AN ESQUIMAUX HUT.

Russia and southern Scandinavia, in Europe, southern Siberia in Asia, and the British Provinces in North America. The

Questions.—519. What does the Warmth Equator show? On which side of the equator does it mostly lie, and where does it attain its greatest distance? 520. Describe the course of the isothermal line of 70°. What does this line mark? 521. Describe the course of the isothermal line of 50°. 522. The isothermal line of 30°. What does this line mark? 523. What is said of the isothermal lines indicating a lower degree of temperature? That of 20°? Of 10°? 524. What is said of the comparative temperature of western Europe and eastern America?

Questions.—What is the first cause mentioned? The second? The third? 525. into what regions may the globe be divided with reference to climate and productions? 526. What are the limits of the Hot Regions? What abound here? What countries does it include? 527. Where do the Warm Regions extend? What districts do they include? What characterizes these regions? 528. What are the bounds of the Temperate Regions? Their temperature? What is said of the seasons here? What countries are included in this region? 529. What do the cold regions include? The seasons here?

summers are short, hot, and oppressive, and the winters severe and protracted. Nearly six months in the year the temperature of Stockholm and St. Petersburg is below the freezing point.

530. The *Frozen Regions* extend from the northern boundary of the oak to the pole. The birch, the hardiest of trees, generally ceases to grow about latitude 70° in Europe, where man is compelled to give up the cultivation of grain. Shrubs and vegetation linger on farther north; grasses and lichens then are only to be met with; and eternal snows and ice succeed. In regions north of 58° in eastern Asia, 71° on the western coast of Europe, and 59° in America, east of the Rocky Mountains, the mean temperature is below the freezing point. Around Hudson Bay and in North Siberia, lakes and standing water of no great depth are frozen to the bottom; the inhabitants remain crowded together in small huts; and if the cold air suddenly enters a habitation, the vapors fall in a shower of snow.

531. Observation in high northern latitudes has pretty nearly determined the fact that the point of greatest cold is not coincident with the pole, but that the lowest temperature is found at two points situated at about 78° of latitude, and 95° west

longitude, and 130° east. Captain Parry, who wintered at Melville Island, often observed the thermometer in the ship at 50°, and at a distance from the ship at 55° below zero. He wintered on the south coast of the island in about latitude 74°, and obtained the following results of observation:

The greatest heat at Melville Island was....+60°   Fahr. on the 17th July.
The greatest cold at   ditto,   —50   "   on the 15th Feb.
Mean temp. of warmest month, July........+42.41   "
    "   of coldest month, Feb. ....—32.19   "
    "   of winter—Dec., Jan., Feb .....—28.02   "
    "   of spring—March, April, May. ..—03.27   "
    "   of summer—June, July, Aug...+37.11   "
    "   of autumn—Sept., Oct., Nov...—00.51   "
The mean temperature for 12 months ......+01.88   "

Dr. Kane observed at his winter quarters, on the northeastern border of Smith's Strait, a temperature, on the 7th of January, 1855, of 69.3° below zero—the severest cold which had ever been noted by strictly scientific observation.

NOTE.—*The map-questions relating to Temperature, on pages 55 and 87—95, should be studied before proceeding with the following lessons.*

Questions.—580. Where do the Frozen Regions extend? Give particulars respecting this region.

Questions.—581. What has observation in high northern latitudes determined? Where are the supposed points of the greatest cold?

PART IV.

THE ORGANIC CREATION.

LESSON I.

PLANTS.

HE term *organic creation* is applied to those objects which have life, and are possessed of parts, or organs. Organized bodies are either animals or plants. With animals, the stomach is the organ of digestion, the lungs of respiration, the limbs of motion, etc. A plant is also composed of parts, as the root, bark, leaves, etc., which severally perform certain functions necessary to its existence and growth.

533. Botanical Geography, or the "Geography of Plants," is a branch of natural science that treats of the laws which regulate the distribution of vegetable life over the surface of the earth. Plants occur over the whole globe under the most opposite conditions. They flourish in the bosom of the ocean as well as on land, under the extremes of cold and heat, in polar and equatorial regions, on the hardest rocks and in the most fertile valleys, amid the perpetual snow of lofty mountains and in springs at the temperature of boiling water, and in deep caverns where the sun has never sent a cheering ray.

534. The life and healthy growth of plants depend upon light, heat, and moisture. It is *light* that gives to plants that beautiful green color, the intensity of which increases with the brilliancy of the light. "Plants always turn toward the light; the guiding power we know not, but the evidence of some impulsive or attracting force is strong, and the purpose for which they are constituted to obey it is proved to be the dependence of vegetable existence upon luminous power."[*]

535. *Heat* is another essential which determines the condition of plants, by the amount of it which prevails during the season of vegetation. In the cold regions of very high latitudes vegetation scarcely exists, and even in the temperate zones its luxuriance is materially diminished by the severe climate of winter. The influence of heat on vegetable life is most strikingly exhibited on high mountains in the torrid zone, where the growth and luxuriance of plants diminish in proportion to elevation, and consequently in proportion to the diminution of heat.

536. Without *moisture* there can be no vegetation, and this element is supplied to the plant in three different forms: that of vapor which the plant absorbs,—that of liquid water by which some plants are surrounded, and that of moisture which the plant extracts, as nutriment from the earth. The development of plants is further dependent upon the chemical qualities of the soil in which they live, whether it be composed of pure or mixed earths, or of soil rich in clay or vegetable mold.

537. It belongs to Botany, Vegetable Physiology, and Agricultural Chemistry to investigate the structure and nature of plants, and to examine in detail the treasures of the vegetable kingdom. The business of the physical geographer is to notice the disposition of the vegetable tribes, and the circumstances which seem to regulate their distribution.

538. Scarcely fourteen hundred species of plants appear to have been known by the Greeks and Romans. At the time of Linnæus (A.D. 1762) the number of known species was 8,800. In 1835, Lindley gave the number at 86,000. At the present time, according to Lyell, there have been collected upward of 100,000 species; and when we reflect that the interior of Africa, of Australia, and of the great islands of Oceanica have not been visited by the naturalist, it will not be deemed extravagant to estimate the total aggregate of species on the earth at 133,000.

539. "A species embraces all such individuals as may have originated from a common stock. Such individuals bear an essential resemblance to each other, as well as to their common parent, in all their parts. Thus the white clover is a *species*, embracing thousands of cotemporary individuals, scattered over our hills and plains, all of a common descent, and producing other individuals of their own kind from their seed."[o]

540. The vegetable kingdom consists of two great natural divisions, namely, Phænogamia, or Flowering Plants, and Cryptogamia, or Flowerless Plants. The PHÆNOGAMIA possess a woody structure, have leafy appendages, develop flowers, and produce seeds. They have two subdivisions, depending upon their manner of growth, called Exogens and Endogens.

541. The *Exogens* (from *exo*, outside, and *genesis*, increase) are a class of flowering plants whose stems have bark, wood, and pith. The bark is increased by layers deposited within the previously formed layers, and the wood by layers or rings placed outside of those of the previous year. This class embraces the forest trees, as the oak, elm, pine, chestnut, poplar, hazel, willow, birch, etc., most of the flowering shrubs and herbs, as the arbutus, sage, mint; also the dahlia, artichoke, thistle, lettuce, marigold, dandelion, daisy, etc. They are also called *Dycotyledons*, from the seed consisting of two lobes.

[*] "The Poetry of Science," by Robert Hunt.

[o] Wood's "Class-Book of Botany."

542. As a new layer is formed every year, it is easy to determine the age of an exogenous tree by counting the number of layers or rings. In this way De Candolle advances proof of the following ages:

Elm	335 years.	Oriental Plane, 720 years and upward.
Cypress, about	350 "	Cedar of Lebanon, about 800 years.
Cheirostemon, about	400 "	Oak	810, 1080, 1500 "
Ivy	450 "	Lime	1076, 1147 "
Larch	576 "	Yew	1214, 1458, 2588, 2880 ys.
Orange	630 "	Taxodium	4000 to 6000 "
Olive	700 "	Baobab	5150 "

543. The *Endogens* (from *endon*, within, and *genesis*, to increase) are those which have their stems increasing from within, and present no separate appearance of wood, pith, and bark. They comprehend the numerous grasses, and the most important of all vegetable tribes, viz., the valuable pasture and all the grain-yielding plants, wheat, barley, oats, Indian corn, rice, sugar-cane, etc., with lilies and the palm family. They are also called *Monocotyledons*, from having only one seed-lobe.

SUGAR-CANE AND RICE.

544. The CRYPTOGAMIA, or flowerless plants, include mosses, lichens, fungi, ferns, sea-weeds, etc.

545. Station indicates the peculiar nature of the locality where each species is accustomed to grow, and has reference to climate, soil, humidity, light, elevation above the sea, etc.; by habitation is meant a general indication of the country where a plant grows wild. Thus the *station* of a plant may be a salt-marsh, a hillside, the bed of the sea, or a stagnant pool. Its *habitation* may be Europe, North America, or New Holland, between the tropics.

LESSON II.

DISTRIBUTION OF PLANTS.

546. IN considering the distribution of the vegetable species it is important to observe the distinction between indigenous and exotic plants. The former are the native productions of a country; the latter are those which have been introduced from abroad. The number of exotic plants is comparatively small. They consist, for the most part, of those species which are eminently useful to man in furnishing him food, the materials for clothing, etc., besides a variety of flowering plants and shrubs.

547. The indigenous class comprehends the great proportion of the vegetable species which adorn the surface of the globe. It includes many useful plants which can not be successfully transplanted to foreign climes; but by far the greater number

are those which are not so especially serviceable to man, and hence there is no inducement to transfer them from the countries in which they are naturally found.

548. Of the indigenous plants, it has been ascertained that different regions are inhabited by distinct species. This fact is strikingly exhibited by an examination of New Holland, where they are found to be, almost without exception, distinct from those known in other parts of the world. Countries situated between the same parallels of latitude differ essentially in their species of vegetation. Out of 2,891 species of flowering plants observed by a naturalist in the United States, there were only 385 which are found in northern or temperate Europe. Humboldt and Bonpland, in all their travels in equinoctial America, found only twenty-four species common to America and any other part of the world.

549. It is a remarkable fact, that in the more widely separated parts of the Eastern Continent, notwithstanding the existence of an uninterrupted land communication, the diversity of species is almost as striking as between countries separated by wide oceans. Thus there is found one assemblage of species in China, another in the countries bordering the Black Sea and the Caspian, a third in those surrounding the Mediterranean, a fourth in the great platforms of Siberia and Tartary, and so forth.

550. By the term *botanical province* is meant a district the vegetation of which consists in great part of species confined to the limits of that district. Twenty-five great botanical provinces have been established, although many of these contain a variety of species which are common to several others. Professor Martius, of Munich, has divided the vegetation of the globe into 51 provinces, namely, 5 in Europe, 11 in Africa, 13 in Asia, 3 in New Holland, 4 in North America, and 8 in South America, besides the province of Central America, the Antilles, the Antarctic Lands, New Zealand, Van Diemen's Land, New Guinea, and Polynesia.

551. "The first travelers were persuaded that they should find, in distant regions, the plants of their own country, and they took a pleasure in giving them the same names. It was some time before this illusion was dissipated; but so fully sensible did botanists at last become of the extreme smallness of the number of phænogamous plants common to different continents, that the ancient floras fell into disrepute. All grew diffident of the pretended identification; and we now find that every naturalist is inclined to examine each supposed exception with scrupulous severity. If they admit the fact, they begin to speculate on the mode whereby the seeds may have been transported from one country to the other, or inquire on which of two continents the plant was indigenous, assuming that a species, like an individual, can not have two birthplaces."[*]

552. Plants are diffused in a variety of ways. The principal of the inanimate agents provided by nature for scattering the seeds of plants over the globe are the movements of the atmosphere and of the ocean, and the constant flow of water from the mountains to the sea. A great number of seeds are furnished with downy and feathery appendages, enabling them, when ripe, to float in the air, and to be wafted easily to great distances by the most gentle breeze. As winds often prevail for days and weeks, or even months together, in the same

Questions.—542. How may the age of an exogenous tree be determined? Ages of several species of trees? 543. What are the Endogens? What do they comprehend? By what other name are they called, and why? 544. What do the Cryptogamia include? 545. What does station indicate? Habitation? Illustrate. 546 What are indigenous plants? Exotic plants? Of what species do the exotics, for the most part, consist? 547. What does the indigenous class comprehend? What does it include? 548. What has been ascertained respecting the indigenous plants?

Questions.—What is said of the vegetation of different countries situated between the same parallels of latitude? Illustrate. 549. What is said of the diversity of species in the Eastern Continent? Illustrate. 550. What is meant by the term botanical province? How many botanical provinces have been established? 552. What are the principal inanimate agents employed in scattering the seeds of plants? Explain the agency of winds.

* Lyell's "Principles of Geology."

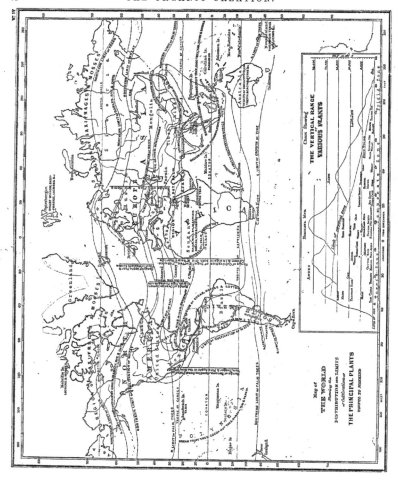

direction, such seeds may be conveyed to a great distance from the parent plant. Even the heavier grains may be borne through considerable spaces by the hurricanes and whirlwinds which prevail in some regions. The germs of many plants, as mosses, fungi, and lichens, consist of a fine powder, the particles of which are scarcely visible to the naked eye, and there is no difficulty in accounting for their being dispersed throughout the atmosphere, and carried to any point of the globe where there is a station fitted for them.

553. Rivers aid in the distribution of vegetation by bringing down to the valleys the seeds which may accidentally fall into them. Thus the southern shores of the Baltic are visited by seeds which grew in the interior of Germany, and the western shores of the Atlantic by seeds that generated in the interior of America. The Gulf Stream is known to convey to the western coasts of Europe the fruits and seeds of plants, which are indigenous to America and the West Indies, in such a state that they might have vegetated had the climate and soil been favorable.

554. Seeds are also distributed by the agency of animals. Some kinds adhere, by means of prickles, hooks, and hairs, with which they are provided, to the coats of animals, to which they remain attached for weeks, or even months, and are borne along into every region whither quadrupeds may migrate.

555. The diffusion of the vegetable species is also promoted by migratory birds, which every year alternate in millions between different countries, and convey to and fro the seeds of plants preserved in their gizzards, or attached to their feathers. When we reflect that these causes have acted incessantly for thousands of years, we can not doubt the immense effect of their joint influence in distributing the forms of vegetable life upon the surface of the globe.

BREAD-FRUIT TREE.

556. But man has been a voluntary agent in effecting the same object, with reference to those plants which are serviceable for food and clothing, or as articles of luxury and ornament, or for building purposes.

Questions.—553. Of rivers. The Gulf Stream. 554. How distributed by animals? 555. By birds? 556. What kinds of plants have been distributed by the voluntary agency of man?

557. "The date-palm has been introduced from Africa into the south of Spain. The grapevine, now so common in western Europe, has been naturalized from western Asia. The coffee bush, native to the highlands of Ethiopia, was taken thence to the scene of its present cultivation. the southern part of the Arabian peninsula; and the culture of the tea-plant, indigenous to China, has recently been attempted with success in the south of France. Rice, known in the southern regions of Asia from the remotest antiquity; the valuable bread-fruit tree, indigenous in the same district, and in the Polynesian Islands; and the more important cereals, wheat, barley, oats, and rye, have all been transferred from the Old World to the New since the discovery of the latter by Europeans."

558. "In return, the Old World has received from the New, maize or Indian corn, and the potato, the cultivation of which extends from Lapland to the extremity of Africa. Our principal fruit trees appear to have traveled into Europe and Great Britain from Syria; the damson plum, with the damask rose, as their names import, from the neighborhood of Damascus; the cherry from Pontus; the walnut and the peach from Persia; the apricot from Armenia; the citron, lemon, and orange from the warmer parts of Asia."○

NOTE.—*The map-questions on page 73 should receive attention before proceeding with the following lesson.*

LESSON III.
FOOD PLANTS.

559. MAN, in almost every country, has selected *annual plants* for food; that is, such plants as complete their whole vegetative processes within the course of a few months. These, for the most part, possess a subterraneous and usually tuberous stem, which sends up shoots above the soil; after some time flowers appear, and afterward fruit. During the remainder of the year the plants sleep, as it were, beneath a protecting coverlet of earth, and are thus beyond the influence of excessive heat or cold. By the cultivation of these plants, man has rendered himself independent of the destroying action of the dry season in semi-tropical regions, and of the killing influence of the winter cold in higher latitudes. It is remarkable that there are only three arborescent vegetables in the whole world which can be included among the chief food-plants, namely, the *bread-fruit*, the *cocoa-nut*, and the *date*, and these have become objects of culture, and furnish in certain regions the principal food of large bodies of men.

560. The most common food-plants are as follows: In the Old World the species which prevail are the *grains*, or *cereal grasses*, namely, *barley*, *oats*, *rye*, *wheat*, *rice*, *millet*, *sorghum*, and the *olive*. The trees are the *date-palm*, *banana*, *cocoa-nut*, *bread-fruit*, and the *pandanus*. In the New World the species which have their origin are *maize*, *potato*, *manioc*, and *arrow-root*. The food-plants cultivated to a certain extent in both continents are *sugar*, *coffee*, *tea*, the *vine*, *cocoa*, *pepper*, *cinnamon*, *cloves*, *nutmegs*, and *cassia*.

561. The regions over which these plants are distributed may be seen by inspection of Map 4. They range from the poles toward the equator in the following order. *Barley*, which has the widest distribution of all the cereals, is cultivated from

Questions.—559. What plants has man selected for his food? What is said of these plants? Which are the only three arborescent vegetables included among the chief food-plants? 560. What species of food-plants prevail in the Old World? The trees?

* Rev. Thomas Milner.

the extreme limits of culture in Lapland, to the heights immediately beneath the equator, but it is only in a narrow zone of the northern hemisphere that it is reared as the sole bread-grain; beside it appear *Rye*, which is the peculiar bread-grain in Norway, Sweden, Finland, and the northern part of Russia; and *Oats*, which are extensively cultivated throughout northern Europe. These grains are also cultivated in North America, though chiefly confined to British America and the northern part of the United States.

562. *Wheat* is the prevailing cultivated plant in Great Britain, Germany, France, and a wide range toward the east, including the whole of the region of the Caspian; in the basin of the Mediterranean, and throughout North America, it is associated with maize. Its northern limit in America is unknown, the country being uninhabited; but at Cumberland House, in the very middle of the continent, one of the stations of the Hudson Bay Company, in 54° north latitude, there are fields of wheat, barley, and Indian corn. Wheat thrives luxuriantly in Chile and Rio de la Plata, and at elevations of 8,500 and 10,000 feet above the sea. It even produces grain on the banks of Lake Titicaca in the Peru-Bolivian Andes, at the absolute height of 12,795 feet, in sheltered situations. *Buckwheat* is a plant of tolerably extensive culture in the northern temperate zone. It is a native of Asia, from whence it was brought into Europe in the fifteenth century.

563. *Maize* or *Indian Corn* is much cultivated in Europe and America south of the 47th parallel of latitude. In the United States great attention is paid to the culture of this grain, of which there were produced, in 1860, over 830 million bushels. *Rice* has been cultivated in the southern regions of Asia from the earliest ages. It constitutes the staple food of the inhabitants of the Indian peninsulas, China, Japan, and the East Indian Islands. Rice is the food of a greater number of human beings than any other grain. It requires excessive moisture, and a temperature of 73° at least; consequently its cultivation is limited to countries between the equator and the 45th parallel.

INDIAN CORN.

564. The *Olive* in the Old World embraces two zones or bands, north and south of the equator, about 9° in width, from latitude 35° to latitude 44°. The climate of the New World, which is subject to the extremes of heat and cold, is not favorable to the cultivation of this plant. The *Date-Palm* yields one of the most nourishing fruits in existence. It grows spontaneously on the southern slopes of the Atlas chain, on the banks of the Nile, and in the Canary Islands; its range extends to Palestine and Hindoostan, and it has been introduced into the south of Spain. It is said that each tree yields annually from 150 to 260 pounds of fruit.

565. The Date-Palm is a member of the palm family, which grows chiefly in the tropics. From their noble and stately appearance, palms have been called by Linnæus "the princes of the vegetable kingdom." Wine, oil, flax, flour, sugar, and salt, says Humboldt, are the produce of this tribe; to which Von Martius adds thread, utensils, weapons, food, and habitations. The cultivation of the date-palm is an object of high importance in the countries of the East. In the interior of Barbary, in Egypt, in the drier districts of Syria, and in Arabia, it is almost the sole object of culture. It is a slow-growing tree, and it has been estimated that the age of one sixty feet high can not be less than 300 years. It is so abundant, and so unmixed with anything else that can be considered as a tree in the country between the States of Barbary and the Desert, that this region is designated the Land of Dates (*Belad-el-Jareed*).

566. The *Banana* or *Plantain*, which yields a great amount of nutritious substance, is a native of the southern portion of the Asiatic continent, but has been transplanted into the Indian Archipelago and Africa, and has also found its way into the tropical parts of the New World. The fruits weigh, altogether, about 70 or 80 pounds, and the same space which will bear 1,000 pounds of potatoes, brings forth, in a much shorter time, 44,000 pounds of bananas; and if we take account of the nutritious matter which this fruit contains, a surface which, sown with wheat, feeds one man, when planted with bananas affords sustenance for twenty-five.

BANANA AND COCOA-NUT TREES.

567. The *Cocoa-nut Palm* also belongs to inter-tropical climes, flourishing especially on islands and near the sea-shores. It is cultivated nowhere so extensively as in the islands of Ceylon, Sumatra, and Java. This tree grows to the average height of eighty feet. The nuts hang in clusters of a dozen each on the top of the tree. The fruit consists externally of strong, tough, stringy filaments resembling coarse oakum, which is formed into *coir*, and extensively used in the East for making cordage. Inclosed within this fibrous mass is the shell, of great firmness, and used for many domestic purposes; While the nut is green, the whole hollow of the shell is filled with an agreeable, sweet-

ish, refreshing liquor. When the nut is gathered, a formation of albumen takes place upon the inside of the shell, producing that white, firm, pleasant-tasted, but rather indigestible substance which is called the kernel of the nut. A tree generally furnishes about 100 nuts.

568. The *Bread-fruit Tree* is distributed generally among the Friendly, the Society, and the Caroline Islands. The tree is beautiful as well as useful, and rises to the height of about forty feet; when full grown it is from a foot to fifteen inches in diameter. The fruit is green, heart-shaped, about nine inches long, and equaling a large melon in size. The nuts, when roasted, are said to be as excellent as the best chestnuts, but it is principally for the fleshy receptacle or pulp that it is valued. When roasted it is soft, tender, and white, resembling the crumb of a loaf, but it must be eaten new or it becomes hard. Others compare the flavor to that of a roasted potato. Such is the abundance of the fruit, that two or three trees will suffice for a man's yearly supply.

569. The *Potato* is a native of Chile, and was first introduced into Britain from Virginia, by Sir Walter Raleigh, in 1586. Of all the vegetable productions especially adapted for the sustenance of man it has the widest range, extending, according to Humboldt, from the northern extremity of Africa to Labrador, Iceland, and Lapland. In tropical regions an elevation of 4,000 feet is necessary for its growth.

570. The *Cassava* or *Manioc* is a shrub, a native of Brazil. The fleshy roots of this plant yield a nutritious substance, from which we obtain our tapioca. This latter article is a kind of starch, and is capable of being made into excellent puddings; it is a very wholesome food for children, and for persons whose digestive organs are feeble. A considerable quantity of this preparation is exported annually from Brazil to the United States. *Arrow-Root*, which forms a pleasant and useful aliment for children and invalids, is much cultivated both for domestic use and for exportation in the West India Islands, Surinam, and in some parts of Hindoostan. It is so named from the property it is said to possess of being an antidote to the poisoned arrows of the Indians.

571. *Sugar-Cane.*—The range of this plant may be said to extend to all the regions of the torrid zone. In countries where the mean temperature is not less than 64°, it extends to latitude 30° on each side of the equator in the New World, and to about latitude 35° or 36° in the Old. The native country of the sugar-cane chiefly cultivated is China; wild sugar-cane was found growing in many parts of America, on the discovery of the New World. The cultivated sugar-cane was conveyed to Arabia, Nubia, Egypt, and Ethiopia, where it became an object of extensive tillage. Early in the fifteenth century it first appeared in Europe. Shortly after the discovery of the New World by Columbus, the plant was conveyed to Hayti and Brazil, from which latter country it gradually spread through the islands of the West Indies and the southern part of the United States.

572. *Coffee.* This bush or tree is a native of the Ethiopian highlands of Africa, whence it has been introduced into Arabia, the East and West Indian Islands, Surinam, Cayenne, and

Questions.—565. What is said of the palm family of plants? What do they produce? What is remarked respecting the cultivation of the date-palm? 566. The banana or plantain? Its productiveness? 567 The cocoa nut palm?

Questions.—568. The bread-fruit tree. Describe the fruit. 569. The potato? 570. The cassava, or manioc? Arrow-root? 571. Sugar-cane? Its native country? Into what regions was the cultivated sugar-cane successively introduced? 572. Coffee?

Brazil. The tree grows upright with a single stem, covered with a light-brown bark, to the height of from eight to twelve feet, and has long, undivided, slender, horizontal branches, which cross each other. These are furnished with evergreen, opposite leaves, not unlike those of the bay tree. The flowers

TEA AND COFFEE FIELDS.

grow in clusters at the root of the leaves, and close to the branches. They are of a pure white, and of an agreeable odor. The fruit, which is a berry, grows in clusters along the branches, under the axils of the leaves. Each berry contains two seeds.

573. Tea consists of the dried leaves of the tea tree, which is indigenous to China and Japan. It was first imported by the Dutch in 1610, and was brought into England in 1666. "The tea country" in China is situated on the eastern side between the 30th and 33d parallel of north latitude. A species of holly, called Paraguay Tea, grows in the forests of Paraguay, and yields a beverage called Maté in Brazil.

574. The Vine comes to perfection in Europe as far north as latitude 50° or 52°, but its profitable culture does not extend much beyond latitude 48°,—the best wines being produced between the 30° and 45° north latitude. Its range in America is much more limited, a difference of 10° occurring between its limits in the Old and in the New World. The Fig is the fruit of a small tree with broad leaves. It is produced in India, Turkey, Greece, France, Spain, Italy, and northern Africa; but the best figs come from Turkey. The Cocoa, or Chocolate tree, grows spontaneously in several of the countries of tropical America; its fruit resembles a cucumber, the seeds of which furnish a substance from which chocolate is prepared.

LESSON IV.

ANIMALS.

575. THE animal kingdom is supposed to comprehend not less than 250,000 distinct species, of which, however, an immense proportion belong to the insect class. Its members are arranged in four grand divisions, namely: 1. Vertebrated animals. 2. Molluscous animals. 3. Articulated animals. 4. Radiated animals.

576. Vertebrated animals are those which have a vertebral column, or back-bone. This department is subdivided into four classes: 1. Mammalia, or animals which suckle their young. 2. Birds. 3. Reptiles. 4. Fishes.

577. Mammalia, or the mammiferous class, stand at the head of the animal creation, and are distributed into eight orders—besides man, who forms a distinct order, termed Bimana (two-handed). Though they differ vastly in appearance and habit, they nevertheless correspond in one particular, that of suckling their young. The eight orders, with some of their types, are as follows: 1. Quadrumana (four-handed),—monkeys, apes. 2. Carnivora (flesh-eaters),—cat, hyena, lion, bear, civet, glutton, mole. 3. Marsupialia (pouched),—opossum, kangaroo, wombat. 4. Rodentia (gnaw-

ers), beaver, porcupine, squirrel. 5. Edentata (toothless),—sloth, armadillo, ornithorynchus. 6. Pachydermata (thick-skinned),—elephant, rhinoceros, hippopotamus, zebra, tapir, horse, ass, and hog. 7. Ruminantia (chewing the cud), —camel, ox, goat, sheep, deer, antelope, giraffe. 8. Cetacea (belonging to whales),—whale, dolphin, narwhal, seal, porpoise.

578. Birds are distributed into the following six orders : 1. Rapaces—Birds of prey ; 2. Scansores—Climbers ; 3. Oscines—Songsters ; 4. Gallinaceæ—Hen-like birds (gallina, a hen) ; 5. Grallatores—Waders ; 6. Natatores—Swimmers. The known number of species is upward of 6,000.

579. Reptiles are distributed into the following four orders, viz : 1. Chelonia (Tortoises) ; 2. Sauria (Lizards) ; 3. Ophidia (Serpents) ; 4. Batrachia (Frogs).

580. Molluscous, or soft-bodied animals, are those which have no bones, but whose muscles are attached to a soft skin, which is inclosed, with few exceptions, in a hard case or shell. In this department there are three classes : 1. Those which have their heads furnished with feet, as the cuttle-fish, nautilus, etc. 2. Those creeping on the stomach,—the slug, snail, limpet, and whelk. 3. Headless, as the oyster, muscle.

581. Articulated animals are those in which the body is divided into joints or rings, sometimes hard and sometimes soft, which supply the place of a skeleton ; this department consists of four classes: 1. Annelides, or ringed worms, as earth-worms and leeches. 2. Crustacea, animals covered with a hard crust, as crabs, lobsters, and shrimps. 3. Insecta, of various families, as flies, bees, wasps, and butterflies.

Questions.—573. Tea? Paraguay tea? 574. The vine? Fig? The cocoa or chocolate tree? 575. How many distinct species is the animal kingdom supposed to comprehend? Into what four grand divisions are its members arranged? 576. What are the vertebrated animals? How is this department subdivided? 577. What is said of the mammalia?

Questions.—Into what eight orders are they divided ? 578. Into what six orders are birds distributed ? 579. Into what four are reptiles ? 580. What are molluscous or soft-bodied animals ? What six classes does this department include ? 581. What are articulated animals ? Of how many, and what, classes does this department consist ?

India and Africa, erect pyramids of clay to the height of ten or twelve feet, sufficiently compact to sustain the weight of several men,—far more wonderful works, in proportion to the size of the builders, than the pyramids of Egypt.

585. No part of the world is so remarkable for the profusion of insect life as the regions of the Orinoco, and other great rivers of tropical America. Humboldt informs us that at no season of the year, at no hour of the day or night, can rest be found there, and that whole districts in the upper Orinoco are deserted on account of these insects. Different species follow one another with such precision that the time of day or night may be known accurately from their humming noise, and from the different sensations of pain which the different poisons produce. The only respite is the interval of a few minutes between the departure of one gang and the arrival of their successors, for the species do not mix. On some parts of the Orinoco the air is one dense cloud of poisonous insects to the height of 20 feet.

586. Among vertebrated animals, the reptiles are especially numerous and formidable in this zone,—as the crocodile of Africa, the gavial of India, and the alligator of America,—and the serpent tribe, some distinguished by their prodigious length

BOA CONSTRICTOR.

and power, the python of India, and the boa of America, and others of smaller proportions, armed with a poison of peculiar deadliness, the hooded snake of Asia, the cerastes of Africa, and the yellow viper of America. The birds here are of the most beautiful forms, splendid colors, and largest dimensions,—as the graceful birds of Paradise, inhabiting New Guinea; the parrot tribe of Brazil, the ostrich of Africa, and the cassowary of Australia. The mammiferous quadrupeds are likewise found, in tropical regions, in the greatest variety, including the most colossal, the elephant, rhinoceros, hippopotamus, and giraffe; and the most sanguinary, the lion, tiger, leopard, panther, ounce, hyena, puma, and jaguar.

587. The animal tribes of the temperate and cold regions are, with a few exceptions, distinguished for their positive utility to man. Advancing from the equator toward the pole, they are found, as a general rule, gradually to diminish in number, magnitude, and ferocity. The insects of temperate regions are much smaller than their tropical fellows, and except in the hottest parts of the year, and in marshy localities, they produce little inconvenience. The great voracious reptiles totally disappear, and the venomous serpents are few and upon a smaller scale. The birds of two orders, swimmers and waders, chiefly subsisting on fish, increase in numbers with distance from the equator; the songsters also have more melodious notes in temperate than in tropical countries; but all the varieties are marked with greater simplicity of coloring.

[CONTINUED ON PAGE 81]

Questions.—585. Profusion of insect life in tropical countries? 586. What is said of the vertebrated animals of tropical regions? Examples? Examples of large mammiferous quadrupeds? 587. For what are the animal tribes of the temperate and cold regions distinguished? The insects of these regions? The reptiles? The birds? What is said of their colors?

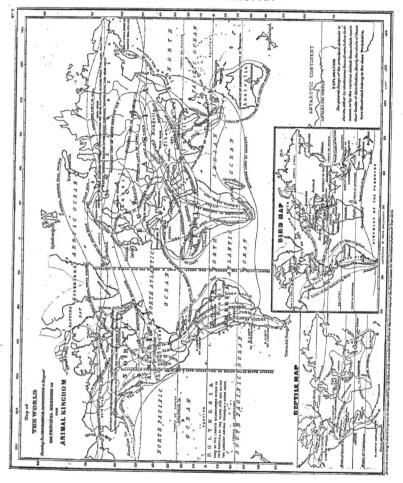

QUESTIONS ON THE MAP.

DISTRIBUTION OF MAMMALIA.

WHAT noted cetaceous animals abound in the arctic seas? What large and powerful bear, inhabiting the frozen regions, has its southern limit mostly north of the Arctic Circle?

What species of ox inhabits the northern lowlands of America and the arctic islands west of Baffin Bay? What kind of deer is found in the southern part of the arctic regions? Does its range also extend through the cold temperate regions? What other member of the deer family has a similar range in the temperate regions, but extending somewhat farther south? Are the reindeer and moose of the New World like those of the Old World? Ans. They resemble them, but are of different species.

The northern portions of the continents form the chief district of what kind of animals? What fierce, rapacious animal, whose northern limit extends in general a little north of the Arctic Circle, ranges through a great part of the temperate regions? What huge species of bear, noted for its ferocity, is confined to the western highlands of temperate North America?

What animal of the ox genus abounds in the interior plains of the same climatic section of America?

What four species of ruminating animals constituting a distinct genus are confined to the western highlands of South America? To what is this genus closely allied? Ans. To the camel genus. Which species is used as a beast of burden? Ans. The llama. What of its size? Ans. It is intermediate between that of the common goat and cow. What two pachydermatous animals inhabit most of South America and part of Central America? Ans. X—s, p—s. What of their size? Ans. The tapir is nearly the same size as the common hog, but somewhat taller; the peccary is much smaller.

In the middle and southern ... regions of what grand divisions does the camel have a ... range? What ... of camel ... most of this section? Ans. The Bactrian. (two-...) ... In what part of the torrid region of Asia is the ... ? What species is found here? A. This one-humped camel or dromedary. ... what part of the temperate and torrid regions of ... is the range of the one-humped camel prolonged? In what grand division do "antelopes of various species" abound? In what part of Asia do they exist? What ... is found in the ... regions of southern Europe, Asia, and ... of Asia southwest of the ... ? Ans. C—s. What animal of the hog genus has a ... range in Europe, Asia, and along the northern border of Africa? Ans. They are believed to be such as result merely from domestication. ... In what grand division do the elephant and rhinoceros have ... their range? Does the elephant range extend as far north in ... as the species of elephant and rhinoceros the ... in both ... ? Are the species of elephant and rhinoceros the ... in both ... ? They are not. In ... of what grand division does the hippopotamus exist? Does its range nearly coincide with that of the elephant of ...? Where does it ... well somewhat farther north? Ans. In the valley of the Nile and that of the Red Sea. What two

animals of the horse genus are confined to nearly the same range as the elephant of Africa? Ans. Z—a, d—w.

What stately ruminating animal is likewise limited to this range? Ans. G—e.

What animals of the horse and ox genera belong to the temperate region of southern Africa?

Does there appear to be an unusual number of large herbivorous animals in Africa?

What fierce carnivorous animal is confined to central and southwestern Asia, including the islands of Sumatra and Java? What two range through southern Asia and northern Africa? Through nearly the whole of what grand division does the range of the lion extend? In what part of Asia is the lion found? What marked difference is observed between the lion of Asia and that of Africa? Ans. The name of the former is scanty; hide that of the latter is exceedingly ..., and has a peculiar majesty.

What part of Asia and Africa does the jackal inhabit? Through what grand division does the range of the hyena family; ...? Through what portions of Asia? What is the only species extending into the ...? Ans. The striped hyena. What live ... small ranges ... South America and the southern half of North America? Ans. P—a. By what other names is the puma known? Ans. ..., panther, and American lion. What of its ... to the panther and lion of the Old World? Ans. It is by no means strongly marked.

What fine carnivorous animal is found in most parts of South America, especially in Brazil? Ans. J—r. By what other name is the jaguar known? Ans. The South American tiger. To what ... kind of the Old World does it bear the most striking resemblance? Ans. The leopard. How do the puma and jaguar compare in size with the lion and tiger of the Old World? Ans. They are much smaller.

Do large animals appear to be as common on the Western Continent as on the Eastern? To which of the continents does the opossum family, extending through the warm temperate and hot regions, belong? To what great island and a few of the neighboring ... are the marsupials of the Eastern Hemisphere limited? For what is this region noted? Ans. For the extraordinary preponderance of this order over other mammalia within its limits, and for the fact that it contains all of the order of marsupials except the opossum family of America. What animals of the order of edentata abound in South America? Ans. A—s, a—t, c—s, s—t. Does the race of marsupials and edentata on the Western Continent (the former unknown on the Eastern; the latter very sparingly ...) ... te in some measure for the ... of larger animals?

Is the range of the monkey tribes limited to the warm regions of the globe? Does it ... im any part of Asia, or Polynesia? What are the most northern ... takes it reaches in the Old World? What is the only ... mer of ... then not ... in its limits? Does the range of the monkey ... takes extend as far north in the New World as in the Old? What part of Central America and the West Indies does it include? Below the ... mth of what great river in South America does it extend? What genus of quadrumana is especially characteristic of Madagascar and vicinity?

In what peninsula and large island of the East Indies is the orang-outang, noted for its resemblance to man, found? In what part of Africa do the species of the orang genus, known as the chimpanzee and gorilla exist? What of the gorilla? Ans. It is believed to be the nearest approach to the human being, among the lower animals; having a striking physical likeness to man,—being larger, except in height,—of prodigious muscular power, and possessing a strong guidled, but ferocious disposition. Are any of the orang genus found in America? Ans. There are not. We have observed that the ruminants, pachyderms, and carnivora of the Old World are, in general, superior to those of the New,—are the quadrumana or monkey tribe etc. of the former also superior to those of the latter? Ans. They are.

What of the quadrupeds of Polynesia?

DISTRIBUTION OF BIRDS.

The arctic coasts and the northern borders of the Atlantic and Pacific form the district of what bird whose down is an important article of commerce? Near the extremities of what peninsula on each side of the Atlantic does the southern limit of cider ducks extend? The southern borders of the Atlantic, Pacific, and Indian oceans are embraced in the district of what antarctic sea-fowl?

The district of what tiny and beautiful bird comprises South America and all of North America except the colder regions? Of how many species does the humming bird family consist? Ans. About 500. To which grand division are all but a few confined? Ans. South America.

What gallinaceous bird, now extensively domesticated, has its native haunt only in North America east of the Rocky Mountains, and from near the latitude of the great lakes to the Isthmus of Panama? What bird is remarkably abundant in the temperate regions of eastern North America? Ans. P—s.

The district of what member of the vulture family, noted as the largest bird of flight, is confined to the Andes region of South America? Below what elevation does the condor seldom descend except in pursuit of prey? Ans. An elevation of nearly two miles above the sea. How high does it soar? Ans. Six times as high as the region of clouds.

What large-bodied sea-fowl is found along the southwestern coast of South America?

Below the northern limit of what remarkable family of birds do the warm temperate and hot regions of North America lie? Near what part of the Western Continent does the southern limit of this family extend? What member of the parrot family, noted for its exceedingly gaudy plumage, belongs to the eastern part of South America? Ans. M—w. What climbing bird, distinguished for the extraordinary size of its beak, inhabits the same region?

The southern half of Europe, with the neighboring parts of Asia and Africa, form the district of what remarkably sweet-singing bird?

The southern half of Asia, except west of the Persian Gulf and south of the Black Sea, is the district of what richly plumaged gallinaceous birds? What are among the most beautiful species of this genus? Ans. The golden pheasant and the argus-eyed pheasant. What elegant gallinaceous bird is a native of Hindostan? Where does the gallinaceous order of birds comprise the greatest number of species? *Ans. In southeastern Asia.*

What two great ——— of Asia are embraced within the —— age of parrots? In —— oly the whole of the East Indi —— and of Australia included —— with this rang —— b What narrow gen —— and —— small extremity are the only parts of —— ica not within the range of —— ? —— of —— ee of the parrot family on the Eastern —— ——.

What bird, ——, —— for its height and for the elegant plies it ——, —— the dry —— of Africa? *Ans. G.* —— of the —— ation of the —— ? It is incapable of flight, but its —— cant it its coursing —— the rapid, and thus ——————. "probably the fleetest of all running ————". Sumatra, Java, and smaller —— of the East India ——, —— the district of what species of ostrich? *Ans. Cy—.* What species —— ? —— ? *Ans. E-w.* Are —— —— of the —— family in —— ? —— There are two species: the —— ostrich and the Patagonian ostrich. Does the —— ostrich outrival in size and elegance the other —— of this family? *Ans. It does.* What aquatic wading-bird, held sacred by the —— ient Egyptians, is —— —— of —— ? *Ans. I-s.*

What gallinaceous fowl has its native haunt in western Africa? What rapacious bird, noted for preying on serpents, inhabits the southern extremity of Africa and some other parts of the same grand division?

What remarkably beautiful birds belong to the island of New Guinea?

What singular species of swan is found in Australia?

In what parts of the world do the feathered tribes display the most gorgeous plumage? *Ans. In the hot and damp regions of South America and the East Indies.* Where do they comprise the greatest —— ber of species? *Ans. In tropical S—th America.* Of —— orders of —— is is the a remarkable variety in —— regi —— b *Ans. Song-birds and climbers (the latter including the parrots ——).* Are the songsters of South America particularly distinguished for their melody? *Ans. They are not.* What —— grad division contains the greatest —— ber of feathered species in the temperate regions? *Ans. Europe.* What orders are more variously pres —— ted here —— in any other part of the world? *Ans. Swimming birds and wading birds.* Of what —— ther order —— Europe contain a —— greg —— ber of species? *Ans. Songsters.* For —— many are many of the songsters of Europe celebrated? *Ans. For the —— ——*

remarkable sweetness of their notes. Where are rapacious birds most numerous? *Ans. In South America and Europe, or, more specifically, where the songsters and other small birds offording an easy prey are most abundant.*

DISTRIBUTION OF REPTILES.

Is —— ny the whole of the arctic z —— ne beyond the northern limit of reptiles? In the valley of —— hat river d —— es this limit : —— nd a —— tile beyond the —— tic Circle? —— That —— ind of reptiles are found in the far —— thern regions? Is the southern point of Greenland within the regi —— n of —— hise? Are —— re any reptiles in Iceland? In the southern —— part of —— hat cold peninsula on the east of Asia is a species of —— bird found? Is the southern limit of reptiles in —— ica as far north of —— Ultra del Ego? What kind of reptile is found near this limit?

The northern —— slit of —— hat —— rder of repti —— s in America is in about the —— tide of the —— mth of —— lames Bay? Of what, near the mouth of the St. Lawrence River?

What venomous snakes belong to the mild and warm climates of middle and eastern North America and to the hot climate of middle and eastern South America? What to the warmer latitudes of the Old World, as far east as China and Farther India?

The northern border of the district of what huge reptiles extends on the eastern coast of America several degrees north of the peninsula of Florida? Does the southern border of this district reach to within a few degrees of the La Plata River? Does the district of crocodiles occupy nearly as wide a range on the western coast of America as on the eastern? Does it ex —— tend as far north as the Gulf of California? Does it reach much south of the equator? What are the crocodile species of the New World commonly termed? *Ans. Alligators or cay-mans.*

Does the district of crocodiles on the western coast of Africa c —— upy nearly the whole breadth of the t —— rrid zone? Does it e —— nd —— ver —— ain half way —— when the —— ator and the —— of Capricorn on the eastern —— ast? Does it embrace n —— rly the whole of the valleys of the Nile and Red Sea? Does it —— rable the Persian Gulf? Does its northern —— slit extend —— nch —— rth of the p —— ninsula of Hindostan and Farther India? Does the district of crocodiles embrace the —— hole of the East Indies? The northwestern coast of what neighboring lang island d —— es it —— nite? To what grand division do the true crocodiles b —— long? *Ans. To Africa.* What are the crocodile —— ies of the tich termed? *Ans. Gavials.*

What kind of turtle, yielding the valuable tortoise-shell of commerce, is found in the region of the Molucca Islands and the northwestern part of New Guinea? In the region of what large islands of America is this kind of turtle also found? In which region does it yield the most valuable shell? *Ans. In the region of the Moluccas and New Guinea.* What huge tortoise (perhaps the largest known species) belongs to the Galapagos Islands, west of South America?

Do the —— tain genus (the —— bers of —— hich are remarkable for —— sir changing —— lors) belong exclusively to the —— ld or New World? To what —— gnd division do they pertain chiefly? In —— hat part of Asia do they —— xist? In what great island are hey —— und?

What genus of reptil spclosely allied to the —— hird genus, —— xphs in the —— mer parts of —— hica? For what are —— me mbers of this genus remarkable? *Ans. For their —— lge size and —— he agreeable food —— hich their flesh affords.*

Are reptiles —— mast abundant on the Eastern or Western —— tinent? *Ans. On the Western.* What —— rder of reptiles is especially predominant on the —— tern —— tih? *Ans. The amphibia or frog —— rder.* In —— hat respect is this part of the globe —— liter suited to the —— pace of a —— tile of reptiles? *tse. In the prevalence of —— ant —— land plains and the great —— tet of marshes and forests, especially in the warm —— pee.* —— des this abundance of reptiles in America seem to make up in part the deficiency of —— mal life —— hich arises from infert rity in its representatives of the higher —— he? In —— hat part of —— rica do —— tiles met —— ht? *Ans. In the hot and moist regions of Brazil. In —— hat part of the Old World are they —— ast abundant? Ans. In the East Indies.* What —— rder, notwithstanding this —— ct, is wholly wanting in the East Indian Archipelago? *Ans. The frog order.* What one is lack —— ing —— but Polyn sia? *Ans. The serpent order.*

CONCLUSIONS.

We have —— —— sh in the foregoing connection that the Old World is superior in the *animal kingdom* to the New, —— with the consideration of a former —— mp [*see page 56*]—that the New World is the —— mre productive or superior in its *vegetable kingdom*; do —— tse —— psige —— tes of the two thus —— ppar to counter —— hak —— teo each —— ther to form a thus —— lete superiority of the New World —— ully —— tated? *Ans. With its —— —*. With —— hat does the animal superiority of the Old World app —— to be associated? *Ans. With its dryness; since it is in the dry re-gions—as in Africa, for instance—that it is most strongly —— hid.*

588. In the temperate and cold regions the quadrupeds of the carnivorous order are chiefly represented by the lynx, wild-cat, weasel, fox, wolf, and bear; the rodents, or gnawing animals, by the porcupine, squirrel, rat, mouse, hare, and beaver; the ruminants, which chew the cud, by various species of the ox, sheep, goat, and deer tribes; and the cetacea, inhabitants of the ocean, by the porpoise, seal, walrus, narwhal, and whale. The herbivorous reindeer and musk ox, and the carnivorous arctic fox and white polar bear, are restricted to the coldest climates, as the herbivorous rhinoceros and elephant and carnivorous tiger and hyena are to the hottest.

LION-SEALS OR SEA-LIONS.

589. The animals of temperate and cold districts are generally remarkable for a tendency to be gregarious or social. Wolves often hunt in packs; beavers form colonies; the wild goats and mountain sheep, with the domestic breeds, associate in flocks; the bison, or American buffalo, is rarely seen solitary on the plains of the Missouri, but has membership with a vast herd; and the reindeer, with other kindred species, have eminently the same characteristics. The gregarious tendency of some kinds of fish is observed in the immense shoals of herrings and mackerel that visit our coasts, and the salmon and shad that flock into our rivers at certain seasons of the year. Many kinds of birds are remarkably gregarious, of which the wild pigeon of America is an example.

590. Animals are adapted to different climates and diverse physical circumstances, by clothing, differing in quantity and quality. The quadrupeds of the torrid zone are mostly furnished with a coat of short and thin hair; but with increasing latitudes, soft and abundant fleeces become common; while in still colder regions, the beaver, sable, ermine, and bear are supplied with the thickest furs. The aquatic tribes of birds which swim in the cold waters of high latitudes are supplied with a compact coat of oily feathers, which abound most upon the breast, as it, in swimming, first meets and cleaves the cold fluid. The right-whale and walrus, which permanently dwell in the cold ocean, derive protection from the chilling waters by the enormous amount of blubber, a bad conductor, which surrounds their bodies.

NOTE.—The map-questions relating to mammalia, birds, reptiles, etc., on pages 79 and 80, should be studied before proceeding with the following lesson.

LESSON VI.

ZOOLOGICAL REGIONS.

591. THE peculiarities which distinguish the animals of the warm, the temperate, and the cold parts of the earth were briefly explained in the preceding lesson. But we have learned, from our examinations of the map, that the same species are by no means universal in countries within the same latitudes or which have the same temperature. Different regions are characterized by distinct species of animals as well as vegetables. It was observed by Buffon that "when America was discovered, its indigenous quadrupeds were all dissimilar to those previously known in the Old World. The elephant, the rhinoceros, the hippopotamus, the cameleopard, the camel, the dromedary, the buffalo (Asiatic), the horse, the ass, the lion, the tiger, the apes, the baboons, and a number of other mammalia, were nowhere to be met with on the new continent; while in the old, the American species of the same great class were nowhere to be seen—the tapir, the llama, the peccary, the jaguar, the cougar, the agouti, the paca, the coati, and the sloth."

592. The earth has been divided by naturalists into ten zoological regions. First. The European region, which comprehends, besides Europe, the borders of the Mediterranean, and even the north of Africa, and extends into Asia, beyond the Ural Mountains and the Caspian. The bear, the fox, the hare, the rabbit, and the deer are among the animals which belong to this district.

NOTE.—It is important to bear in mind that the same name is often applied, in different countries, to animals of different species. Thus in North America are found bears, foxes, rabbits, and deer; but they are of a different species from those of Europe. The buffaloes which roam in vast herds over the prairies of America are very unlike those of India; and the species of India and America differ greatly from the Cape buffalo of southern Africa.

593. Secondly. The African region is inhabited by many animals not found elsewhere. The hippopotamus, for example, the giraffe, the zebra, the chimpanzee and gorilla, and the thumbless apes, are exclusively African. A few of the species inhabiting the northern confines of this continent, such as the dromedary and jackal, are common to Asia. The elephant of Africa is smaller, has a rounder head and much larger ears than the Indian one, and has only three instead of four nails on each hind foot. In like manner, not one of the four African species of rhinoceros agrees with either of the three Indian kinds.

594. Thirdly. The southern region of Africa, where that continent extends into the temperate zone, constitutes another separate zoological province. This region is cut off from the countries of milder climate in the northern hemisphere by the intervening torrid zone. Here may be found peculiar species of the rhinoceros, the hog, and the hyrax, among the thick-skinned races; and among the ruminating, the Cape buffalo, and a variety of remarkable antelopes, as the springbok, the oryx, the gnou, and several others.

595. Fourthly. The island of Madagascar, though separated from Africa by a channel only 300 miles wide, forms another

province, all the species, except one, being peculiar. This district is distinguished by the number of the Lemur genus (akin to monkeys) which inhabit it.

596. *Fifthly.* Another of the great nations of terrestrial mammalia is that of *India*, containing a great variety of peculiar forms, such as the sloth-bear, the musk-deer, the nylghau, the gibbon or long-armed ape, and many others. *Sixthly.* A portion of the *Indian Archipelago*, embracing the large islands of Java, Sumatra, and Borneo. A few of the species inhabiting these islands are common to the continent of India, but most of them are distinct.

597. *Seventhly.* The islands of Celebes, Amboina, and *New Guinea* constitute another region, in which are found many marsupial quadrupeds. Of this region Lyell remarks: "As we proceed in a southwesterly direction, from Celebes to Amboina and thence to New Guinea, we find the Indian types diminishing in number, and the Australian (*i. e.*, marsupial forms) increasing. Thus in New Guinea seven species of pouched quadrupeds have been detected, and among them two singular tree-kangaroos; yet only one species of the whole seven, viz., the flying opossum, is common to the Indian Archipelago and the mainland of Australia."

598. *Eighthly.* When *Australia* was discovered, its land quadrupeds belonged almost exclusively to the marsupial or pouched tribe, such as the kangaroos, wombats, flying opossums, kangaroo-rats, and others. From recent investigations it has been ascertained that there are no less than 170 species of marsupial quadrupeds, and of the whole number all but thirty-two are exclusively restricted to Australia.

599. *Ninthly.* North America constitutes another vast zoological province, inhabited by species of animals very rarely identical with those of Europe or Asia. The influence of climate in limiting the range of animals is here conspicuously

BUFFALO HUNT.

displayed. The animals of the State of New York are of quite a different species from those of the arctic regions, and also from those of South Carolina and Georgia. Among the quadrupeds which inhabit the northern part of the continent are the musk-ox, polar bear, and the caribou or American reindeer, besides numerous species of fur-bearing animals. The grizzly

bear, the largest and most ferocious of its kind, inhabits the western highlands; and the American buffalo roams in herds of thousands over the prairies which extend west from the head waters of the Mississippi.

600. *Tenthly.* South America is the most distinct, with the exception of Australia, of all the provinces into which the mammalia can be classed geographically. The monkey tribe, which are very numerous in the forests of Brazil, differ widely from those in the Old World; many of them having prehensile tails, and all being noted for their widely separated nostrils. The sloths and armadilloes, the true blood-sucking bats or vampires, and many other animals, are peculiar to South America.

LESSON VII.

MAN.

601. THE number of the human race is variously estimated at from 1,000 to 1,200 millions. The impossibility of stating it with any degree of precision arises from the fact, that in but few countries is any enumeration of the inhabitants ever made; while there are vast and populous regions which have scarcely been visited by civilized man. The best opinion appears to be that 1,200 millions is a close approximation to the real number.

602. Mankind, pre-eminently distinguished from the lower animals by the high endowments of reason, conscience, and speech, also differ from them in consisting only of a single species. They are geographically diffused through almost every climate, from the hottest to the most frigid. Under the scorching rays of a tropical sun, upon the banks of the Senegal, the human body supports a heat which causes alcohol to boil; and in the polar regions of northeast Asia it resists a cold which freezes quicksilver.

603. Few lands have been discovered destitute of a native human population. Iceland, Spitzbergen, Nova Zembla, Madeira, the Azores, St. Helena, the Falkland Isles, and South Shetlands, some groups in the Pacific Ocean, as the Galapagos Isles, and a large number of small islets, with all the lands within the Antarctic Circle, were uninhabited when first made known to European nations, but have, in several instances, since been colonized. The range of man extends from the 75th parallel of north latitude to the 55th of south latitude. The most northern dwellers are the Esquimaux, on the shores of Baffin Bay; the most southern are the inhabitants of Tierra del Fuego.

604. On man's capability of accommodating himself to a great diversity of circumstances, Dr. Paley remarks: "The human animal is the only one which is naked,[o] and the only one which can clothe itself. This is one of the properties which renders him an animal of all climates and of all seasons. He can adapt the warmth or lightness of his covering to the temperature of his habitation. Had he been born with a fleece upon his back, although he might have been comforted by its warmth in high latitudes, it would have oppressed him by its weight and heat as the species spread toward the equator.

605. "Within the tropics, where a vegetable diet is found to be most

Questions.—596. What forms are peculiar to India? What is said of the species inhabiting the Indian Archipelago? 597. Of the islands of Celebes, Amboina, and New Guinea? What does Lyell remark of this region? 598. What is said of the land quadrupeds inhabiting Australia on its discovery? 599. What is said of the animals of North America? Of the influence of climate in limiting the range of animals? Name some of the principal quadrupeds of this division. 600. What is said of South America? Illustrate. 601. How is the number of the human race variously estimated?

Questions.—Most probable number? 602. How are mankind distinguished from the lower animals? What is said of their wide diffusion? 603. What regions were uninhabited when first made known to Europeans? Between what parallels does the range of man extend? 604, 605. Give the substance of Dr. Paley's remarks.

• This remark, though not strictly true, is applicable in respect to nearly all the higher animals.

grateful and conducive to health, nature supplies, in the greatest abundance, the most valuable vegetable productions. In the temperate regions, animal food is more or less abundant; and the various kinds of grain, roots,

NATIVES OF TIERRA DEL FUEGO.

and fruit afford plentiful and wholesome nutriment. As we approach the polar regions, grains and fruit gradually disappear, and animal food becomes more and more exclusively used, until we reach the Samoyeds and Esquimaux, who are unacquainted with bread.''

LESSON VIII.

RACES OF MEN.

606. WE are informed in the Sacred Scriptures that it pleased the Almighty Creator to make of one blood all the nations of the earth, and that all mankind are the offspring of common parents. Though differing greatly in form, stature, features, and complexion, the members of the human race are found to possess no specific differences,—the hideous Esquimaux, the refined and intellectual Caucasian, the thick-lipped Negro, and the fair, blue-eyed Scandinavian being mere varieties of the same species.

607. Classifications of mankind have been based upon the differences that exist in respect to the color of the skin, hair, and eyes, and of the form of the skull. Taking the *color of the hair* as the leading characteristic, there are three principal varieties :—*First.* The *Melanic,** or black class, which includes all individuals or races which have black hair. *Second.* The *Xanthous,*† or fair class, comprising those who have brown, auburn, flaxen, or red hair. *Third.* The *Albino,*‡ or white variety, comprising those whose hair is pure white, and who have also red eyes.

608. Taking the *shape of the skull* as the basis of a classification, mankind are divided into five grand classes or races—the Caucasian, Mongolian, Ethiopic, American, and Malay.

609. In the *Caucasian* race, the head is commonly of the most symmetrical shape, almost round or somewhat oblong; the forehead of moderate extent; the cheek-bones rather narrow, without any projection; the face straight and oval, with the features tolerably distinct; the nose narrow, with the bridge slightly arched; the mouth small, and the lips a little turned out, especially the lower one; and the chin full and rounded.

610. The most perfect examples of this variety are found in the regions of western Asia, bordering on Europe, which skirt the southern foot of the Caucasus Mountains, from whence the class derives its name, and which is near what is supposed to be the parent spot of the human race. Here are the Circassians and Georgians, among whom are found the most exquisite models of female beauty.

611· The Caucasian race, of pure blood, comprises nearly all the ancient and modern inhabitants of Europe, except the Finnian and Samoyedian tribes in the north, the Tartar and Turkish stock in the south, the Magyars or Hungarians in the central section, and the Lettons in the vicinity of the Baltic—all of which varieties are more or less mixed with Mongolian blood. This race also includes most of the inhabitants of southwestern and southern Asia as far as the Brahmapootra River, and of the inhabitants of Africa west of the Red Sea and north of the southern borders of Sahara, together with a great number of Europeans and their descendants who have settled in other parts of the world.

612. In the *Mongolian* race the hair is coarse, straight, and black, the eyes generally rise in an oblique line from the nose to the temples, the arches of the eyebrows are scarcely to be perceived, and the face is broad and flat, with the parts imperfectly distinguished. The complexion is generally of a tawny or olive color, which is described as intermediate between that of wheat and dried orange peel, varying from a tawny white to a swarthy or dusky yellow.

613. This division embraces the tribes that occupy the north, central, east, and southeast parts of Asia: the natives of Siberia, the Chinese Empire, Japan, Farther India, Boothia, and Toorkistan; also a liberal share in the mixed varieties of Europe, previously mentioned,—especially those in the north and east; and the Esquimaux on the shores of the Arctic Ocean and Hudson Bay. A portion of this family

Questions.—606. What do the Sacred Scriptures inform us? 607. Upon what have classifications been based ?

* MELANIC, from the Greek *melan,* black. † XANTHOUS, from the Greek *xanthos,* yellow. ‡ ALBINO, from the Latin *albus,* white.

is distinguished for a considerable degree of culture, especially the Chinese and Japanese, but owing to their exclusive social system, which has separated them from the rest of mankind, they have made but little progress for ages.

614. The *Ethiopic* race have black eyes, black woolly hair, flat noses, thick lips, and a projecting upper jaw. The forehead is retreating, and the head less globular than that of the European. The best examples of this race are the negroes south of the Sahara; in Soudan and Upper and Lower Guinea. The natives of Senegambia and the Kaffres of the southeastern part of Africa resemble others of this race in their jet-black color, and some of their features, but they are taller, more slender, and better proportioned than the rest. ·

615. The nations commonly classed with this race are widely dispersed; they occupy all Africa south of the Great Desert and Abyssinia, Australasia, and some of the interior portions of the larger East India Islands and the Malay Peninsula. . To this race belong also the negroes in America, who were originally brought from Africa, and who have multiplied in the New World to a vast extent.

616. The *American* race are distinguished by a copper-colored complexion, long, coarse, coal-black hair (which is never crisped like that of the African, or curled, as that of the white sometimes is), prominent cheek bones, broad face, and a scantiness of beard.

Their senses of sight, hearing, and smell are remarkably acute. In war and the chase they are indefatigable, but they are averse to regular and mechanical labor. They are cold and phlegmatic in temperament, and manifest an extraordinary insensibility to bodily pain.

617. The native American tribes and nations, excepting the Esquimaux, belong to this class. The Indian tribes of North America are fast disappearing before the spread of the white man, being now confined principally to the unsettled regions west of the Mississippi. Among the most warlike of these tribes are the Sioux and Camanches. The Indians of South America, except those of the Peru-Bolivian stock, are mostly in an abject condition, indolent, and destitute of that nerve and spirit which is known to distinguish their brethren of the north.

618. In the *Malay* class the top of the head is slightly narrowed, the face is wider than that of the negro; the features are generally more prominent; the hair is black; the color of the skin is tawny, but sometimes approaching to that of

mahogany. The division embraces the principal tribes of the Indian Archipelago, and all the islanders of the Pacific, excepting those which belong to the Ethiopic varieties.

619. "The diffusion of mankind over the globe has transpired in the course of ages under the influence of various causes. The pressure of population in one district outstripping the means of subsistence, the love of enterprise, the spirit of acquisition, social disturbances, and foreign violence, have contributed to scatter the human family far from the common center where the race originated. Endowed with intelligence to devise means of surmounting natural barriers—mountains, deserts, rivers, lakes, and the ocean —there is no difficulty in accounting for the geographical range of man. The contiguity of the mainland of northeastern Asia to that of northwestern America, with the nearly-connected chains of the Japan, Kurile, and Aleutian isles intermediate, point to the New World as having received its original population from the Old in that direction. In modern times, adverse winds have driven Japanese junks across from one continent to the other; and, probably, more frequently than we are apt to imagine, crews have been compelled to expatriation by the tempest, surviving its perils, and colonizing distant isles and archipelagoes."—*Rev. Thomas Milner.*

QUESTIONS ON THE MAP.

HABITABLE LIMIT AND SUBSISTENCE OF MAN.

Where is the most northern permanent habitation of man ? What settlement exists there ? Ans. Port Folk, an Esquimaux settlement. What is its latitude. Ans. About 78° 18′ north. How far north do the Samoiedes, in Asia, have a permanent abode ? Does the permanent habitation of man extend farther south than the islands near the southern extremity of the Western Continent ? In what zone do the inhabitants subsist almost wholly on animal diet ? [See perpendicular line near the middle of the map, marked with the kinds of diet in the different zones.] In what one, on little else than vegetable diet ? In what two zones is their diet of mixed animal and vegetable food ?

DISTRIBUTION OF RACES, ETC.

To which of the principal races of mankind do the occupants of the arctic regions belong almost exclusively ? Does this race overspread the entire northern borders of both the continents ? What race occupies the greatest area on the Eastern Continent ? In what peninsula does it extend farthest south ? What are the inhabitants of Farther India called ? What are the principal population on the north of the region occupied by the pure Chinese ? What stock inhabit most of central Asia ? By what other name is it commonly known ? Ans. The Tartar stock. Around the borders of what sea is a limited portion of Europe occupied by the Turkish variety of the Mongolian race ? What branch of this race is the most widely spread in northern Europe ? What stock occupies the northern part of the Scandinavian peninsula ? What is that branch of the Mongolian race which inhabits the northern regions of the Western Hemisphere called ?

Which of the races is most widely dispersed in the mild and warm parts of the world ? Which occupies nearly all southwestern and southern Asia as far as the Brahmapootra River and Himalaya and Hindoo Koosh mountains ? Does this race possess all Europe except the small part inhabited by Mongolians ? Is it the chief race in Africa north of the Tropic of Cancer ? In the basins of what river and sea does it extend considerably south of this section ? What remote extremity of Africa does it occupy ? The southeastern and southwestern borders of what great island does it inhabit ? What large islands in the vicinity does it wholly or partly overspread ?

The exterior portions of what grand division, except in the south, are

*Questio s.—*614. Describe the Ethiopic race. Where are the best examples found ? 615. What is said of their wide dispersion ?

*Questions.—*616. Describe the American race. 617. What tribes and nations belong to this race ? 618. Describe the Malay race. What tribes does this race include ?

Map of
THE WORLD,
showing the
distribution of the principal
RACES AND VARIETIES
OF MANKIND.

EXPLANATION
OF COLORS
European—
Mongolian—
American—
African—
Malay—

mostly occupied by the Caucasian race? What is the only extensive region in the southern half of North America which is not occupied by this race?

Into what two great branches is the Caucasian race divided? *Ans. The Japhetic or Indo-European branch, and the Semitic or Syro-Arabian branch.* Why are these branches termed Japhetic and Semitic? *Ans. Because, by many writers, the members of the former are supposed to be descendants of Japheth, the youngest son of Noah; and those of the latter, of Shem, the oldest son of Noah.*

Which of these branches occupies northern Africa? What great peninsula in Asia, and the country between the river Euphrates and the eastern part of what sea, does it inhabit? On the southern border of what peninsula in Europe is it found? Does this branch predominate elsewhere?

What variety of the Semitic branch is found in Egypt? Of what ancient people are they the present representatives? *Ans. The ancient Egyptians.* Of what is the population of the Barbary States mainly composed? What inhabitants are found on the northwestern border of the Sahara? *Ans. B-s.* What branch of the Caucasian race includes all the families of the same that are widely dispersed? What variety of this branch occupies eastern Europe? What, southwestern Europe? What, the rest of Europe, inhabited by the Caucasian race, including Iceland and most of the British Isles?

In what region of North America are the inhabitants of Indo-European stock largely intermixed with aboriginal American? What part of South America is a continuation of this region? By what Europeans was the southern part of this region mainly settled? *Ans. By Spaniards.* In what part of South America is there a similar region? Between the mouths of what great rivers does it extend? By whom was it chiefly settled? *Ans. By Portuguese.* To what leading variety of the Indo-European branch do both the Spanish and Portuguese belong? *Ans. To the Celtic.*

In what country of South America are the inhabitants of purer Indo-European stock? From whom are they chiefly descended? *Ans. From English,* French, and Dutch. From whom are the Indo-European inhabitants of the United States and provinces on the north descended? *Ans. From the British (chiefly English), and, to a much less extent, from the Germans.* To which of the leading varieties of the Indo-European branch do the population, thus composed, mostly belong? *Ans. To the Teutonic.* Has the Teutonic variety in America ever intermixed much with the aborigines? *Ans. It has not.* Does it present, in this respect, a marked contrast with the Celtic?

Does the true African or negro race constitute the ruling population in any region of considerable extent besides Africa? Have the representatives of this race, in modern times, become voluntary settlers in distant lands? *Ans. They have not.* Would the lack of a commercial character, arising from the limited shore-line of Africa, as compared with its area [*see page 8*], have a tendency to restrict them to their aboriginal home? Would the Sahara form an important barrier to their overland migrations?

What peculiar varieties of this race are found in South Africa?

What variety somewhat resembling those just mentioned is found in New Guinea and the small islands extending thence southeasterly to the Fejee group? What variety constitutes the aboriginal population of Australia?

What island is the most western in the Old World, inhabited by Malays? With what other race are the Malays of this island intermixed? *Ans. The Negro.* What peninsula is the only continental region occupied by Malays? Does this race extend over all Oceanica, except the islands inhabited by the Papuan and Australian varieties? Do the Polynesians, in the eastern part of Oceanica, differ somewhat from the true Malay type? *Ans. They do.* What other race do they resemble? *Ans. The American or Indian race.*

Is the Indian population of America mostly confined to the interior? In what part of North America does it reach the coast? Of South America? What variety of race inhabits the southern part of South America? For what is this variety remarkable? *Ans. For exceeding in stature any other of the varieties of mankind, the males averaging about six feet in height.*

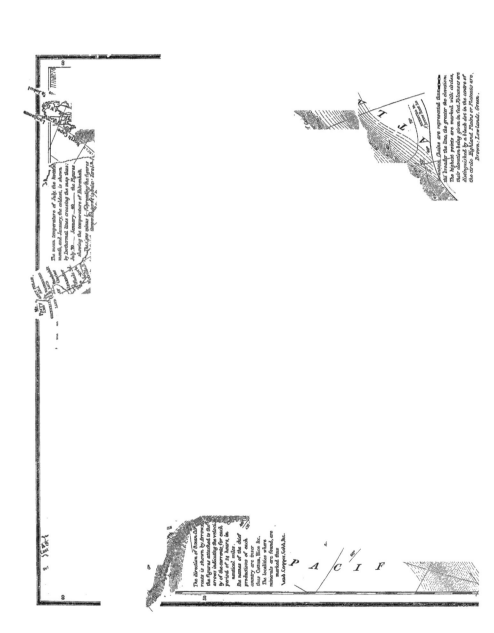

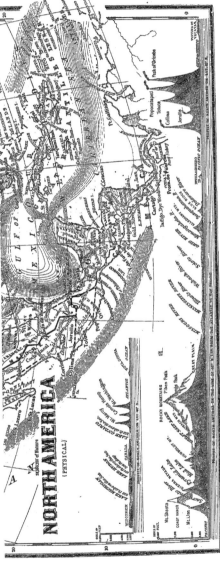

NORTH AMERICA
(PHYSICAL)

QUESTIONS ON MAP OF NORTH AMERICA.

PLATEAUS AND PLAINS.

What part of North America consists almost wholly of highlands? What name has been given to the high-lands on the eastern side of the Rocky Mountains? What region, chiefly desert during most of the year, is embraced in the middle and southern part of this plateau? What similar region lies south of the Great American Desert, or forms a continuation of the same? What is the meaning of the name Llano Estacado? *Ans. Staked plains; so called from the way-marks fixed in the surface for the guidance of travelers.* Are the above-named plateaus shattered on the west by the broader and more elevated parts of the Rocky Mountains?

By what name is the vast plateau-region or succession of table-lands west of the Rocky Mountains designated? What plateau forms the southern continuation of this? What three plateaus lie farther south? What desert extends northward from the Gulf of California? What desolate waste in the eastern part of the Plateau of Anahuac?

What plateau between the valley of the St. Lawrence River and the Atlantic Ocean? *Ans. Laurentian plateau.* By the valleys of what lake and rivers is the continuity of this plateau somewhat broken? *Ans. Lake Champlain and H-n-R.* What plateau forms the basis of a large part of the Alleghany Mountains?

How much of North America consists of lowlands? Name the several lowland sections bordering the great bodies of water on the northeast and southeast. What extensive lake and river basin extends through the central and eastern part of the North American lowlands? What vast plain extends through the heart of the grand division? The lowlands of what two great rivers form the chief part of the plain?

TEMPERATURE.

Do the January isotherms crossing the northern portion of North America bend far south in the interior? Which part, then, embraces most land which is subject to the coldest winters? [It will be borne in mind that the bending of any isotherm to the poleward side of the spot upon the parallel of latitude indicates that a region of which it passes through which it passes]. Are the ——— lands has a colder temperature than any through which it passes.]

coldest parts of the interior the farthest removed from the warm maritime influences of the Atlantic and Pacific? Why do not the Arctic Ocean and Hudson Bay have much influence in lessening the cold in these parts? *Ans. Because, during most of the year, they are thickly sheeted with ice.* In winter the Great Central Plain is swept chiefly by north and northwest winds,—what must be the influence of these winds upon its temperature at that season?

In the Gulf Stream, which overspreads a great part of the North Atlantic, a warm or cold current? How does the January temperature of Iceland—the southern shores of which are bathed by the Gulf Stream—compare with that of Nova Scotia, 20° farther south? As the Gulf Stream, throughout the winter, is free from permanent ice, even as far north as Spitzbergen, must the winds which sweep across it at this season have great influence in lessening the cold of the lands over which they subsequently pass? Easterly and southeasterly winds, during winter, often blow from the Atlantic towards Greenland,—what must be their influence on the temperature of the western part which borders the west coast of Greenland, south of the Arctic Circle, compare with that of Newfoundland and New Brunswick? As northwesterly winds during winter prevail chiefly in the eastern part of North America, is the temperature of Newfoundland and New Brunswick, at this season, much or little influenced by the severe continental climate of the interior of the grand division? Are the currents from the north which wander and flow between the Gulf Stream and the eastern coast of America warm or cold? Are they of much extent and importance than the temperature of the neighboring coast probably overbalanced for the most part by that of the latter? *Ans. It is.* Would it be much or little felt during the prevalence of northeast winds?

The January temperature of what great peninsula on the Pacific coast of America is as warm as that of the vicinity of New York city? About how much does the difference in the latitude of these two regions? What mountains shield the Pacific coast from cold arctic winds? Northwesterly and westerly winds prevail on this coast,—do they impart to the adjacent lands a maritime or inland climate? In the Japan Stream, which flows from the coast of Asia across the North Pacific and along the neighboring coast of

America, a warm or cold current? Does its temperature, owing to the direction of the above-mentioned winds, have much or little influence on the climate of the adjacent parts of America? Much of the warmth of this current is of course lost in passing through its higher latitudes,—is the temperature of its apparent continuation, known as the California Coast Current, above or below that of the neighboring sea? What of the summer temperature of the California Coast Current? *Ans. It is said to be colder than its under-temperature.* To what may this be ascribed? *Ans. Probably to the partial upsinking, at this section, of the cold under-current, whose waters supply the place of those which flow off in the Equatorial Current—the latter current, during summer, originating somewhat farther north. Does the California Current, like that which flows off in the Equatorial Current, exert much influence, owing to the direction of the prevailing winds, on the climate of the neighboring land? Is this influence more marked in summer, or in winter? As the Pacific coast is warmed at the north by the neighboring currents, and cooled at the south (especially during summer), is the difference in the temperature of these two sections of coast much less than it otherwise would be? From what point south of San Francisco to opposite what island a little south of Mount Fairweather does the July southern line extend along the border of the coast? Through what extent of latitude along this coast, then, is the mean July temperature nearly uniform? *Ans. Through more than 20° of latitude, or over 1,000 miles.*

Do the isothermal lines bend far south in crossing the most elevated parts of the Rocky Mountains and adjacent highlands? What does this indicate concerning the influence of these regions? How do the midsummer and midwinter temperatures of Santa Fé compare with those of New York, 90 farther north in the same continent, of the January isotherm of New York? *Ans. Those of the July isotherm a little below it.* What is the isotherm along Santa Fé, and of the July isotherm a little above it? Should we infer from the dryness of the climate in some of the western parts of North America that these parts are subject to severe extremes of temperature? How does the temperature of the southern part of the Colorado Desert compare with that of the district noted for its intense heat, which includes central Arabia and the southeastern part of the Desert of Sahara? What part of the remarkable extreme of heat has been observed here? *Ans. 116° above zero*

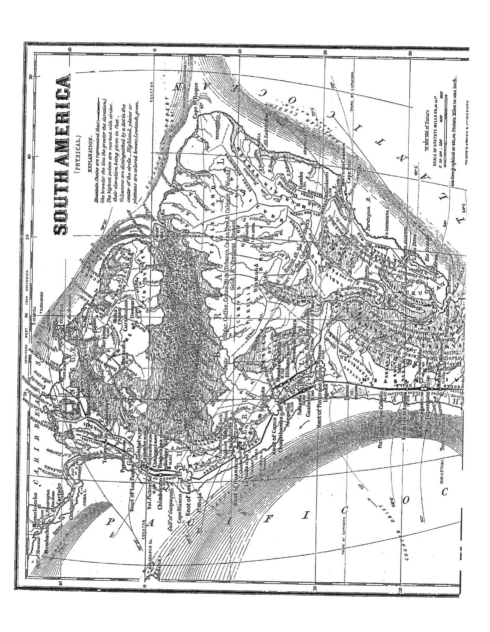

QUESTIONS ON MAP OF SOUTH AMERICA.

PLATEAUS AND PLAINS.

In South America mostly highland or lowland? Are the highlands and lowlands more, or less, interspersed with each other than those of ... th America?

What general ... ame is appli d to the vast plateau region which forms the ... ain of the Andes? *Ans. The | ... ins of the Andes.* How d es it ... ... pare in ... ... ith the plateau region in the ... ern part of North America? What fundamental contrast do these two regions present? *Ans. The plateau region of North America forms the basis of ... the great parallel mountain chains, or is associated with two principal lines of elevation, and occupies the ... hole breadth e ... in them, besides extending along their other ... ... t ; while that of South America forms the basin and highland borders of but one chief ... ... in ... ain.*

... two ... ... es ... ar the middle of the ... Aes region are the ... ... odd ... r... , of large area, throughout its extent? That let ... ... gut by the ... des from the ... Ens of Despoblado? ... ... ma? What ... ne in the ... ab, on the ... ern ide of the ... | ... main system in the n ... rth is the ax is of a ; ... ... le plateau regi o? What is the northeastern part of this re- gi n called? *Plt ... ... ide table-land in the ... ern part of South America? What is the ... tern portion of this ... inland called? Ans. C-s P--s.* The southern portion? Meaning of the name, Campos de Vacaria? *Ans. Cattle fields, so called, because of the immense herds to which they afford pasturage.*

How large a part of South America consists of lowland plains? *Ans. Nearly one third.* What plain or llanos in the north? What one stretches nearly across the grand division, on the southern border of the equator? What large plain or pampas south of this? In what river valley is it embraced? The valley of what great river and its branches in the southw s ...

... plains? What llanos in the northern part of this re- gi o? What barren ... den ... ... cis southwest from thence? *Ans. E. Gn. C-o.* What is that part of the plain which lies nearer to the Rio de la Bita ... ciil? Is the ... tarn half of the grand division sm th of the N gro River lowland or highland?

TEMPERATURE.

Do the isothermal lines drawn across South America extend farther from the equator in the eastern or western part of the grand division? Which part, then, is the warmer? Would the difference in the elevation of the eastern and western parts tend to produce this effect? In which part do we find the loftiest elevations? Are the ocean currents on the eastern side of South America of warm or cold water? *[The currents are distinguished as warm or cold, accordingly as their temperature is higher or lower than that of the neighboring waters.]* Would they render the climate of the neighboring part of the grand division, during the prevalence of winds from the Atlantic, warmer than it otherwise would be? Is the current on the western side of South America, for the most part, warm or cold? What must be its influence, there-

fore, on the climates of the neighboring land during the prevalence of winds from the Pacific? During the cool months the chilling air of this current, along the Peruvian coast, condenses the vapors rising from the neighboring waters to a thick mist, which is borne inland by the sea breezes,—what must be the influence of this mist upon he temperature of the land?

As westerly winds prevail chiefly in the ... uth temperate zone, must we inf e that the cold current, just ... uid, has a greater influence on the climate of the ... tlern part of South America than the Brazilian Current? Is the Ape Horn ... tent warm or ... ld, and what, therefore, is its ... 1 ... nflce on the temperature of the neighboring land?

The ice-fields of the ... ttic regions extend much farther toward the tropics ... han ... ides of the arctic regions,—must their chilling ... he ... fe, during ... e ... ple of southerly ... ws, be felt with ... glar severity in Patagonia? Must we ... far from the ... ode that he ... tern part of South America is subject to : ... las of cold? Does ... taly freezing ... tr ... ais our at ... ra del Fuego in ... ht? *Ans. It does.*

What ... st de the effect of the immense ... fei which covers the plain of the Amazon, on the temperature of the equatorial regi n of South ... al? ... lit ... he fact that ... ... ly the whole of South America ... ... an ... ke ... ... is is open to the ... sd winds from the Atlantic and is abundantly refreshed with periodical rains, give to its torrid regions a cooler climate than they would otherwise possess?

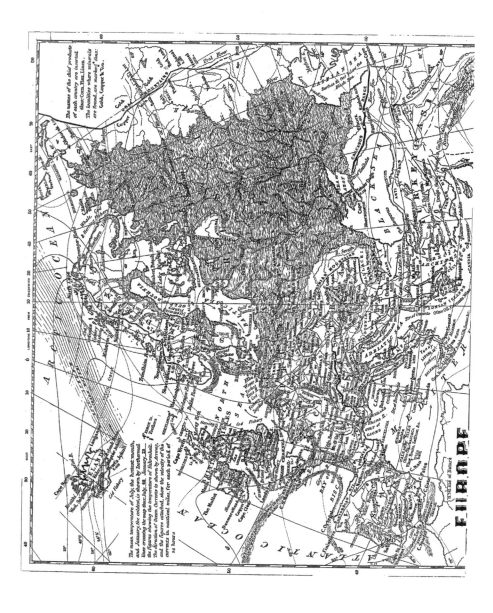

EUROPE

QUESTIONS ON MAP OF EUROPE.

PLATEAUS AND PLAINS.

In what part of Europe is the principal highland region?

What plateau of unbroken area, in the southwest, is the most elevated and extensive in Europe? *Ans. The Spanish Plateau.* What small, uneven plateau north of the Bernese Alps? *Ans. The Plateau of Switzerland.* What plateau, in central Europe, northeast of the last?

What great peninsula in the northwest consists chiefly of highlands?

What part of Europe is mostly a vast lowland plain? Of what does this plain form the western portion? *Ans. Of the great northern plain of the Eastern Continent.* What are the marshy flats along its northern borders? What plain occupies the middle portion of the great lowland of Europe? What plains in the southeast? What one inclosed westward from the Black Sea? What plain extends in the same river valley, northwest of the last? What one nearly in the middle of Europe? What plain south of the Baltic and North Sea? East of the Bay of Biscay and English Channel? What one extends northwest from the head of the Adriatic?

TEMPERATURE.

Do the January isotherms, intersecting Europe, extend farther north in the eastern or western part? Which part, then, is the warmer?

Is the Gulf Stream—the waters of which bathe a great part of the western coast—a warm or cold current? Must the prevailing southwest winds of the north temperate zone bear considerable warmth from this current into western Europe? Does the fact that the warmth of this current is such as to prevent the formation of ice in winter between the northwestern coast of Europe on the one side, and the vicinities of Iceland and Spitzbergen [see map on page 54] on the other, render the winters of the Scandinavian Peninsula much milder? How does the January temperature at North Cape compare with that at the mouth of the river Don, nearly 25° farther south? How does that of the Crimea compare with that of Iceland? During winter, cold north winds prevail in Europe,—would they sweep without obstruction from the Arctic Ocean to the Black Sea? Would their effect, therefore, on the climate at the mouth of the Don and the Crimea be very severe? What shields the Mediterranean countries from their influence? Is the position of Africa such that the diffusion of heat from the Sahara must render these countries warmer than they would be if there were a broad ocean on the south? Is the Rennell Current, which flows along the southwestern coast of Europe, of a character to render the neighboring countries warmer or colder?

Is the nearness of Asia—with its extensive marshy Tundra, and its vast elevated plateaus—conducive to the warmth or coolness of eastern Europe?

Is the eastern or western part of Europe the more open to the influences of the sea, and therefore the less subject to extremes of temperature? Do the July isotherms intersecting Europe, in general, extend much farther north on the eastern than on the western borders? Which, then, has the warmer summer temperature? We have before observed that its winter temperature is much colder than that of the former,—must the difference, therefore, between its extremes of heat and cold be very great, as compared with those of western Europe? Does the proximity of the vast continental region of Asia render the climate of eastern Europe much more excessive than it would be if Asia were submerged beneath the ocean?

ASIA

(PHYSICAL)

EXPLANATION

Mountain Chains are represented thus.⎯
the broader the line the greater the elevation;
The highest points are marked with circles:
thus⊙ and the elevation given in feet. Volcanoes
thus⊙ Table lands and Plateaus colored Brown,
Plains, and Low lands are colored Green.
The names of the chief products of each country
are inserted thus: Silk, Tea. The localities where
minerals are found are marked thus: Gold, Silver.

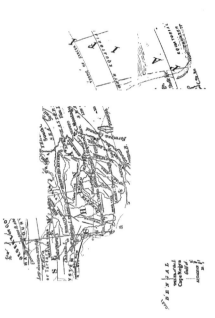

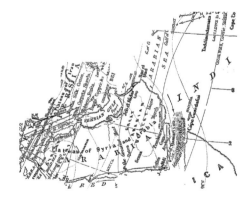

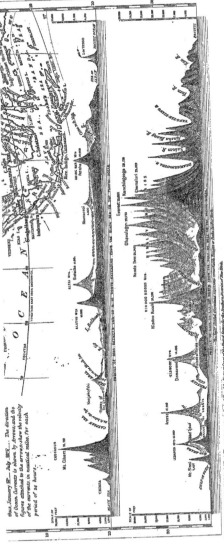

thus, January 2⁰—July 5ᴺᴱ... The direction of Ocean Currents is shown, by arrows; and the figures attached to the arrows, show the velocity of the currents in nautical miles, for each period of 24 hours.

QUESTIONS ON MAP OF ASIA.

PLATEAUS AND PLAINS.

How large a part of Asia consists of highlands? *Ans. Nearly three fourths.* In what three groups are the principal plateaus comprised? *Ans. The plateaus of central and eastern Asia; the plateaus of Hindoostan, and the plateaus of western Asia.*

With how many and what great mountain systems is the central and eastern plateau-region of western North America? Is it more or less remarkable for breadth than the great plateau-region of western North America? What plateau north of the Peninsula of Corea? What one between the Altai Mountain System and the eastern part of the Thian Shan System? What extensive desert-basin on the southern border of the Mongolian Plateau? What plateau south of the Thian Shan Mountains? Between the Kuen Lun Mountains and south of the Bolor Mountains? West of the Bolor Mountains? What familiar name has been applied to this plateau in allusion to its great height and the steepness of its slope toward the neighboring plain? *Ans. "The Roof of the World."*

What plateau occupies most of the Peninsula of Hindoostan? What one on the north, separated from the former by a low chain of mountains?

What considerable desert in the country north of the Arabian Sea? What broad plateau between the Caspian Sea and Persian Gulf? What smaller one between the Aral Sea and Caspian? Between the Black Sea and the Mediterranean? *Ans. The Plateau of Ayotdekia.* What one east of the last? *Ans. The Plateau of Armenia.* What extensive plateau occupies most of southwestern Asia? What desert in the northern part of this plateau? In the southern part? What plain occupies a great part of northern Asia?

What is the waste of marshes and lakes which skirts the Arctic Ocean called? What are the dry grassy tracts in the southwestern part of the Plain of Siberia called? *Ans. Steppes.* What plain adjoins the Caspian and Aral seas? What irregular branching plain in the eastern part of Asia?

What plain between the Himalaya Mountains and the Deccan? What one extends from near the Himalaya to the northeastern coast of the Arabian Sea?

TEMPERATURE.

Around the northern half of what large river in Asia is the "District of the Greatest Cold in January?" How much below zero is the average temperature in that month? [See the figures at the extremity of the lines inclosing it.] What of this temperature, as indicated by the mercurial thermometer? *Ans. It is the lowest thus indicated, the mercury becoming frozen on reaching this point.* In the District of Greatest Cold longest from north to south or from east to west? What of the surface of that part of the Arctic Ocean on the north? *Ans. It is a region of fixed ice—broken in summer, only near the shore [see map, page 56].* Does the above district compare that part of the arctic region of Asia which is farthest removed from the influences of the open Arctic Sea overspread by the waters of the Gulf Stream, and from the influences of the Pacific and the open waters communicating with it through Behring Strait? What part of this district is rendered colder by its highland character? Are the extremes of heat and cold greater in the middle of Siberia than at the east or west? [The bending of the January isotherms toward the equator indicates greater cold.] Of the July isotherms from the equator, greater heat.] What part of Asia has nearly the same January temperature as North Cape in Europe? [See isotherm of 14°.] Do the winter isotherms drawn across Asia extend much farther south in the inte-

rior of the grand division than on the border of Europe? Which part, then, is the colder? Are there vast elevated plateaus in the interior which would tend to produce this difference? Do the July isotherms drawn through the interior of Asia generally extend as far north as on the border of Europe? Are the summers, then, as warm in the former part, notwithstanding its elevation, as in the latter? We have before noticed that the winters in the interior are very severe, the differences between the extremes of heat and cold, therefore, must be excessive.—Is the remoteness of inner Asia from the sea such as would contribute greatly to this excessiveness? We have learned from a former map that but little rain falls in the interior of Asia,—accordingly that it has an arid soil, a dry atmosphere, and an unusually clear sky, —must these facts contribute to its extremes of heat and cold?

Are there both warm and cold ocean-currents on the east of Asia? As the prevailing winds in the temperate region are from the southwest, must we infer that these currents have comparatively little influence on the climate of Asia? Do the northeast winds, which are common in spring and autumn on the eastern coast, pass over a considerable distance along the currents which flow nearest the shore? What, therefore, must be their effect on the neighboring portion of the mainland? Should we infer, therefore, that the climate of eastern China, like that of the eastern part of the United States, is somewhat variable, or subject to unpleasant changes? What must be the effect of the Japan Stream, especially on the climate of Niphon and the neighboring islands on the south?

In what part of Asia is the "District of Greatest Heat in July?" Of what is this district a part? *Ans. Of the region of the greatest summer heat on the globe.* We have before learned that this district lies mostly beyond the range of the monsoons of the Indian Ocean, and is a nearly rainless region,—do these facts render it subject to greater heat?

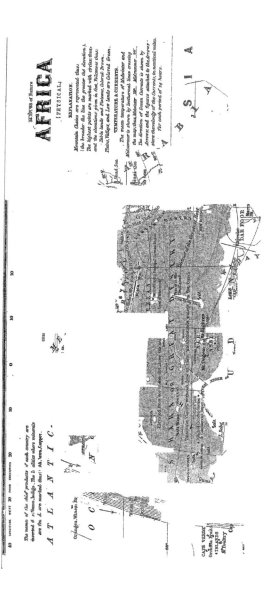

QUESTIONS ON MAP OF AFRICA.

PLATEAUS AND PLAINS.

Is the plateau region of Africa limited to any particular part of the grand divisions, or does it extend through nearly the ——?

What vast —— it occupies —— of the ——
—— in of the —— region? What —— it forms
the —— part of the Sahara? What is the atmospheric ——
ern part called? *Ans. S—t.* What —— of —— and
tween the —— border of Africa? Between —— on the —— and the Red Sea? Below the —— valley of the —— and the Gulf of —— and
—— high —— in —— of the Gulf of —— and
the —— in —— of the Red Sea.

Where is the —— extends in the —— part of Africa? A —— of what desert is included in this region? What —— —— it from the Gulf of —— lowland? *Ans. —— of B–d and J–d.* What great —— river in —— it has its —— —— through
a —— —— lake in the interior is —— —— —— by a —— the lowland and basin? The —— the —— —— river farther west is —— part —— of —— by a low plain? To what parts of Africa are the remaining lands chiefly confined?

TEMPERATURE.

Which is subject to the greatest —— of heat,
the —— in or —— in —— part of Africa? [*See iso-therms of* 0.5°, 8°, *and* 81 9] What arid —— is the chief —— of this —— ? Is —— —— in Africa the ——
the —— or —— part of —— —— pm to the —— ? Which part is the —— ? Which of the cooling influence of —— nic —— of the Nile countries is wholly inclosed in the district of the greatest heat in July which prevails in my part
of the world? What is the mean July temperature
of the border of this district? [*See figures at the end of the line —— they it.*]

In July, the —— appears as far —— as the Tropic of Cancer,—is it more nearly vertical at —— the rate
on the —— —— of Africa, or at the —— ? Do the isotherms indicate —— the —— —— nature —— perature at Cape Bon is higher or lower than at Cape Lopez? The —— ——ins of —— country
in North America have nearly the same —— or —— ?
The United States. Do —— —— the —— —— the law a lower —— ? If, then, the —— or in —— in Af. —— —— also —— hotter in north-

ern Africa than in the lower —— , which has the more variable climate throughout the year?

Is the general outline of Africa which as —— is it —— , or little, —— to the influence of the sea? Is Should we infer from this that its climate on the whole is —— ime or subject to greater —— changes than are usual in the latitudes? We have learned from a former map that Africa, in general, is poorly supplied with rain,—does the —— —— dryness of
its —— , as well as of the soil, and the serenity of its skies, also combine to —— its climate
even ——?

Do the —— —— —— farther from the —— on the —— or —— —— side of south Africa ?
With —— side, —— , is the warmer? What of the temp —— of the Indian —— ? *Ans. It is s-everal degrees higher on that of the other —— .* As —— and —— its prevail chiefly on the western border of the Indian O—— , does its —— the —— influence on the climate of —— Africa?
Is the Moz —— bique —— —— warm or cold? Is the South Atlantic Current of warm or cold —— ? Is its —— with nearly equal to that of the —— ? —— *Ans. It is not.*

P A C I F I C   O C E A N

Strait
Juan de...
Cape Flatte...
M

Grays Harb...
C.Disapointmen...
Pt.Adam...
ASTO

C.Foulweath...

Mc...

Waters

Bay or
San Francisco
SAN FRANCISCO

Mon...
at city...

50°

Santa B.
Bay
Christo B.
a
M
NEWILLE   G U L

UNITED
STATE...
(PHY

SCALE OF STATUT
0   50   100   2

200 Ge

QUESTIONS ON THE MAP.

SITUATION AND EXTENT.

What is the latitude of the most northern limit of the United States? Of the south point of Florida Keys, the most southern limit of the United States? Ans. About 24° 30'. Is this country longest from east to west in the northern or southern part? What is the longitude of the eastern extremity of Maine, the most eastern point of the United States? Ans. About 67° W. l n. Of Cape Flattery, the most western point?

MOUNTAINS AND RIVER SYSTEMS.

What great chain of mountains forms the dividing line between the waters tributary to the Atlantic and those tributary to the Pacific? Is much the greater part of the country on the eastern or western side of this chain? What noted river divides the eastern section into two nearly equal parts? What mountains form the watershed between the valley of this river and the Atlantic? Which are the two highest chains between the Rocky Mountains and the Pacific? Name the three loftiest peaks of the Rocky Mountains in this country, so far as known. What towering summit at the southern extremity of the Cascade Range?

Into how many and what principal slopes and basins is that part of the United States east of the Rocky Mountains divided with respect to its drainage? [*The extent of the principal basins and slopes is indicated by the coloring.*] The part west of the Rocky Mountains? What basin comprises nearly half the whole country? Which is the principal river of the Hudson Bay Basin in the north? Does the Lawrentian Basin—or the valley of the St. Lawrence and Great Lakes—include a very extensive area in the United States? In what direction is the prevailing inclination of the Atlantic slope, or the general course of its rivers, north of Chesapeake Bay? Between the head of Chesapeake Bay and Florida? Of the Gulf slope from Florida to the Mississippi? West of the Mississippi? What great river of the latter section drains the valleys between the southern branches of the Rocky Mountains?

What one drains the Pacific slope south and east of the Great Basin? What is its prevailing course, and into what does it flow? What considerable lake receives the drainage of the northeastern part of the Great Basin? Which is the chief river in the western part of this basin? What river drains the northern part and what one the southern part, of an extensive valley between the Sierra Nevada and the Coast Range? Do both unite in the same outlet? What great river receives the drainage of nearly the whole of the Pacific slope north of the Great Basin?

What is the true parent stream or chief tributary of the Mississippi? Is it longer than the Mississippi as commonly-designated? Are several of the tributaries of the Mississippi-Missouri among the longest rivers in North America?

TEMPERATURE.

What is the mean annual temperature of the northern part of Maine? [*See the isotherms intersecting it.*] Through the northern part of what State west of the Great Lakes does the isotherm of 40° pass? To the northern border of what Territory does this isotherm extend southward into the region of the Rocky Mountains? Should we infer that that part of this region inclosed by the isotherm of 40° embraces the coldest sections of the United States? What gives these sections so low a temperature?

Do the isotherms, generally speaking, bend somewhat southward in crossing the Appalachian Mountain System? Why do they extend less to the south, than in crossing the Rocky Mountains?

What is the mean temperature of the northwestern extremity of the United States? How does it compare with that of the northeastern extremity? Should we infer from its latitude that it would be warmer or colder than the latter? What town in New Mexico has nearly the same mean annual temperature? Is the elevation of Santa Fé more or less than a mile and a quarter (6,830 feet) above the level of the sea? Does the proximity of neighboring mountains, as well as its own elevation, render its climate colder? What city nearly 400 miles farther north on the eastern coast of the United States has the same average temperature?

What is the mean temperature of the northern borders of the Gulf of Mexico? Is it warmer than any part of the Pacific coast of this country?

RAIN.

[*For answers to the following questions, see the corner chart.*] Does more rain fall in the eastern or western part of the United States? Which part borders on the sea to much the greater extent? What large bodies of water in the north contribute to the supply of rain in this section? Does the eastern part of the country have most rain in the north or south? From what gulf is a large part of the vapor derived which supplies the rain in the south? .

Is nearly the whole western section of the United States, from a north and south line crossing the middle course of the Missouri River to near the Pacific, scantily supplied with rain, as compared with the eastern section? What parts of Texas and New Mexico have only ten inches a year, or about one fourth as much as the eastern section? What amount of rain falls over a considerable belt extending from this region to about the middle course of the Missouri? What desolate tra t is comprised in this belt? Ans. The Great American Desert. Does the western half of the plateau region embraced between the Rocky Mountains on one side and the Blue Mountains and Sierra Nevada on the other, have more or less rain than the Great American Desert? What third of this section has only five inches a year? Where is there a small tract which has only three inches a year? What complete desert is comprised in this tract? Ans. The Colorado Desert.

Is the western border of the United States from the vicinity of San Francisco Bay, southward, scantily supplied with rain? What part of the Pacific coast has more rain than any other portion of the United States?

MEXICO   Charlotte   Tampas   C.R

EXPLANATIONS

The average annual temperature is shown by isothermal lines crossing the map ___ 78° ___
Mountain Chains, are represented thus ———————
the broader the line, the greater the elevation .
The highest points are marked with circles.
The figures attached to the names of mountains Passes, Towns, &c. indicate their elevation above tide water in feet. The coloring shows the principal island basins & maritime slopes; as the Mississippi Basin, the Atlantic Slope &c. The names of the chief productions of each country are inserted thus Sugar, Hemp &c. The localities where metals are found are marked thus Gold, Copper.

PHYSICAL GEOGRAPHY OF THE UNITED STATES.

SITUATION AND EXTENT.

The United States, stretching from the Atlantic to the Pacific, and from 49° north latitude to 24° 30′ south latitude, comprise nearly the whole of the mild and warm temperate regions of North America, and thus embrace the most attractive portions of the grand division. The greatest length of the country, from north to south, is nearly 2,400 miles; the greatest breadth, from east to west, about 1,700 miles. Its area is very nearly 3,000,000 square miles, or but little less than that of the entire surface of Europe.

SEA-COASTS, ETC.

In external form this country is remarkably compact, presenting no great peninsula except that of Florida, and being penetrated far inland by no great arm of the sea. Owing, however, to the presence of the Gulf of Mexico on the south, and the oblique directions of the Atlantic and Pacific shores, it has an ample development of coast, and enjoys abundant maritime advantages.

The Atlantic coast and Gulf coast are varied by numerous small inlets and by several bays of considerable extent; so that the total length of the former, including its windings, is nearly 6,900 miles, and of the latter nearly 3,500 miles. The Pacific coast, on the contrary, is more uniform; the mountain ranges which run parallel to it, or extend along its borders—in many parts trenching close on the sea, and presenting a bluff, iron-bound shore; while only at two points do they open for the access of far-invading waters, the narrow inlets of Puget Sound and San Francisco Bay. The whole length of the Pacific coast, therefore, is but little over 2,300 miles, or only two thirds as great as that of the Gulf coast.

On the north the United States have no sea-coast; but nearly equivalent to this are the shores of the Great Lakes, which, although remarkably uniform—affording few harbors except at the mouths of the rivers—have a total length (including their insular lines) of about 3,600 miles.

Thus the entire shores of this country, bordered by great bodies of water, are not far from 16,000 miles in length, or about four fifths as long as the winding sea-coast of Europe, the most maritime of the grand divisions of the globe.

NAVIGABLE INLAND WATERS.

But notwithstanding the compact figure of the United States, or the fact that none of the great bodies of water which surround it penetrate far toward the heart of the country,—its natural facilities for water communication are by no means chiefly confined to its borders. Nor in scarcely any part, except the southwest, is it very difficult of internal access without the aid of artificial channels.

The Mississippi River, with its numerous tributaries ramifying like a vast arterial system through the interior, opens not less than 15,000 miles of navigable water-ways The main stream of the Missouri-Mississippi may be ascended in light-draft steamers from the Gulf of Mexico to the Great Falls of the Missouri, or nearly to the foot of the Rocky Mountains. On the Pacific side of the mountains, the Columbia River may be ascended in like manner to the Cascades; so that the distance between the heads of steamboat navigation on the two streams is only 450 miles.

The Ohio River—the great eastern tributary of the Mississippi—is open to steam navigation as far as Pittsburg, in western Pennsylvania, or to within 200 miles of the head of natural navigation on the Potomac River.

Accordingly, one may pass through the heart of the country, from the Atlantic or Gulf of Mexico to the Pacific, without an overland journey of more than a few hundred miles.

The same easy natural means of communication appear in other directions. The journey from the Gulf of Mexico to the Great Lakes may be accomplished in steamers by way of the Mississippi and Illinois rivers to within a hundred miles of the head of Lake Michigan. In like manner one may pass from the remotest extremity of the Great Lakes to the Atlantic by way of the St. Lawrence and other connecting streams, aided only at a few points by canals to circumvent the rapids.

Easy communication also may be had between New York—the great commercial metropolis of the country—and the St. Lawrence, by way of the Hudson River, joined by a short canal to Lake Champlain and its outlet, the Sorelle.

The number of navigable channels extending inland from the coast to distances varying from near a hundred miles to 800 or more—as the lower courses of the Delaware River, the James, Cape Fear, Savannah, Appalachicola, Alabama, Colorado, and Rio Grande—are too numerous to be mentioned.

On the Pacific borders of the country, however, the extent of navigable streams is quite limited, owing to the near approach of the mountains to the coast, and the fact that they seldom open for the outlet of large rivers from the interior.

GENERAL SURFACE OF THE COUNTRY.

Passing from east to west, the surface of the United States presents eight principal sections, viz., the lowlands which border the Atlantic and Gulf of Mexico, the Appalachian Mountain System and its associate plateaus, the lowlands of the Great Central Plain, the Western Steppes, the Rocky Mountains, the Great Western Plateau, the Pacific maritime chain of mountains, and the Pacific coast region.

The Atlantic lowlands vary in breadth from 50 to 200 or 250 miles.

In the New England States they are seldom more than 100 miles wide, and, except in southeastern Massachusetts, generally present a somewhat hilly surface, diversified with numerous lakes and ponds. The abundance of these sheets of water in some parts of Maine and New Hampshire renders the scenery remarkably picturesque, the country as viewed from the lofty hill-tops seeming literally begemmed with their mirror-like surfaces. The prevailing unevenness of the maritime region of New England renders the coast bold and irregular, and causes the presence of many rocky islands along its border. It abounds, therefore, in deep and well-sheltered harbors which afford superior facilities for commercial intercourse. This character is especially marked on the coast of Maine.

From the Hudson River, southward, the Atlantic lowlands are divided into two principal sections: one, a low plain, bordering the sea, and bounded on the west by a rocky terrace known as "The Ridge;" the other, a hilly region, gradually increasing in ruggedness toward the Alleghanies.

The surface of the plain is wavy in the interior, but flat and frequently marshy over a considerable breadth along the seaboard. Owing to the flatness of the latter section, it has gently sloping and comparatively regular shores, bordered by shallow waters, and affording but few good harbors. Being of an alluvial character, moreover, it is skirted to a great extent by long low islands formed of the accumulated sediment deposited between the spreading currents of the rivers and the inflowing tides of the ocean. In some parts these islands afford but few inlets to the narrow bays or channels which separate them from the mainland. Shifting sand-bars, also, frequently obstruct the inlets or the mouths of the rivers beyond; thus rendering access to the latter comparatively difficult.

The width of the maritime plain varies from fifty to one hundred miles, except in the northern part of New Jersey, where it is much narrower. The larger rivers crossing this plain are generally navigable to the foot of "The Ridge," over which the waters descend in abrupt falls or rapids. These falls are among the most important geographical features of this part of the United States; since, by arresting the progress of navigation and supplying extensive water-power, they have given rise, in their vicinities, to a chain of large commercial and manufacturing towns and cities—such as Newark, Trenton, Baltimore, Georgetown, Richmond, Petersburg, Raleigh, Augusta, and Macon,—which extends, on the line of "The Ridge," through seven of the Atlantic States.

The Peninsula of Florida, lying between the Atlantic and Gulf of Mexico, is wholly embraced in the maritime plain, and to a great extent is flat and marshy.

The Gulf Lowlands are merely a continuation of the Atlantic Lowlands. East of the basin of the Mississippi they are divided, like the latter, into a hilly interior section and a level or wavy coastwise section by a somewhat ill-defined prolongation of the Ridge, through Georgia and Alabama, into Mississippi. Their limit on the north is quite irregular, being formed by the extremities of the plateau and higher ranges of the Appalachian System and by the watershed which separates the basin of the Tombigbee River from that of the Tennessee. The breadth of these lowlands is seldom less than 250 miles, or more than 300 miles. Their most distinguishing characteristics are the well-marked valleys which furrow the hilly regions at intervals from north to south, or from northeast to southwest, and the high ranges of bluffs which border the upper courses of several of the rivers.

West of the basin of the Mississippi the Gulf Lowlands are almost wholly occupied by the maritime plain. They vary in breadth from 150 to nearly 800 miles, and, except a flat coastwise belt covering about one fifth of this breadth, are an extensive undulating prairie.

The Appalachian Mountain System consists of two principal sections, separated from each other by a well-marked depression in the basin of the Hudson River and vicinity, and gradually rising thence to their culminating point in the north and south.

The northern section is the most irregular, embracing, besides its principal ranges and groups, numerous peaks scattered over a wide area, especially in Maine and New Hampshire. It reaches its greatest elevation in the wild and romantic group of the White Mountains, the sublime scenery of which is familiar to travelers from all parts of the world.

The plateau, or broken series of table-lands which belongs to this section, has a rugged outline, and presents but little area except of a hilly or mountainous character.

The southern section of the Appalachian System consists of two parallel zones of elevation: the eastern zone, composed of successive ranges rising one after the other toward the interior; the western, divided from the former by a broad plateau-like valley, and consisting of adjacent table-lands or high plains, wrinkled here and there with longitudinal ranges of hills.

In the northern half of this section, most of the ranges are "remarkable for their regularity, their parallelism, their abrupt acclivities, the almost complete uniformity of their summits, and their moderate elevation. They present the appearance of long and continuous walls, the blue summits of which trace along the horizon a uniform line seldom varied by any peaks or crags."

Advancing southward, however, their elevation increases, the ranges become indented, irregular, and more numerous; until, in North Carolina, "the form of simple parallel ridges" disappears, and, in place of this, we have a broad mountain-tract, thickly and often irregularly studded with lofty peaks. Here, in all respects, is the culminating region of the Appalachian System. Scores of summits rise to a height of 6,000 feet; while the loftiest—Black Dome or Mitchell's High Peak, in the group of the Black Mountains—has an altitude of 6,711 feet, being the highest point of land east of the Mississippi River.

South of North Carolina, the mountains gradually sink into low hills, and the ranges terminate in northern Georgia and Alabama.

The Plateau of the Alleghanies, or southern plateau of the Appalachian System, is traversed by ranges of hills running parallel with the mountain chain of which it forms the basis. Between the Blue Ridge on the east and the Alleghanies and Cumberland Mountains on the west, it consists of a spacious valley, embracing level tracts of considerable area, which extends through Pennsylvania, Maryland, Virginia, and Tennessee. The average elevation of the plateau is scarcely 1,800 feet, but in southern Virginia it reaches a height of 2,000 feet and upward.

The lowlands of the Great Central Plain, in the eastern part, are somewhat broken by the descending terraces of the Appalachian highlands and by the low ranges of mountains or hills which border the same. West of this, they present a vast level or undulating tract, reaching far beyond the Mississippi.

This almost unvarying expanse is made up in great part of half wooded and open prairies—the latter, or true prairies, being most common in Illinois and in the States lying west of the Mississippi, and reaching as far south as the Ozark Mountains. In many sections, however, the landscape is agreeably varied by woody bottom-lands bordering the rivers, and bounded on each side generally by steep acclivities or lines of bluffs.

Along many of the larger rivers the bottom-lands are of considerable breadth. Those of the Missouri, for instance, are five miles wide. But the bottom-lands of the Mississippi below its junction with the Missouri are most remarkable for breadth, being frequently from forty to fifty miles wide; while the bluffs which border them are in some places two or three hundred feet high.

Here and there the marginal flats are penetrated by creeks or bayous, and interspersed with swamps or lagoons. Winding through them, the rivers pursue a variable and uncertain course; the force of their currents, when deviated by slight obstacles, often wearing away the soft alluvial banks until an entire new channel has been gained. Thus, in some places, the whole breadth of the bottom-land is brought on one side of the stream; while, on the opposite side, the river runs close to the bluffs, or even undermines them.

The lower Mississippi is especially noted for its crooked and variable course. Some of its bends, where the intervening neck is only one mile across, are thirty miles around. Channels called cut-offs have been opened across several of these, making an important saving in the navigable distance of the river.

But although the general character of the Lowlands of the Great Plain is that of a level or undulating tract, they include several regions besides that along the eastern border, which present a marked variety of surface. Among these, the most uneven is the hilly tract, embracing the Ozark Mountains, which extend from the interior of Missouri into the north-west corner of Arkansas, and the neighboring border of Indian Territory. Much of Minnesota, also, with the northern part of Wisconsin and the peninsula between Lake Superior and Lake Michigan are comparatively rugged. The scenery of these northern regions is rendered peculiarly attractive by the myriads of lakes embosomed in their valleys, and the numerous water-falls which break the peaceful currents of their rivers.

The Western Steppes, embracing the table-land section of the Great Central Plain, consists mainly of a succession of terraces, generally presenting abrupt edges to the east, and rising one above the other, with gentle slopes, to the Rocky Mountains.

In many places, especially at the south, the rivers crossing these terraces flow through cañons, or deep and narrow defiles, which at the borders of the terraces, in some instances, extend several hundred feet below the general surface.

The mean height of the Steppes, between the Arkansas River and the upper Missouri, is about 4,000 feet; but, along the base of the Rocky Mountains, from the Llano Estacado to the upper Missouri it is not less than 5,000 feet; or, in many parts, more than one mile.

The Rocky Mountains, entering the United States from British America, gradually expand east and west, until the chain attains a breadth of from 200 to 800 miles. The mountain region is made up partly of distinct ranges, and partly of peaks irregularly disposed, frequently with broad level valleys between. The height of the mountains, except in New Mexico, often exceeds 10,000 feet above the sea, and in several instances 14,000 feet; but owing to the elevation of the table-land region that forms their basis, and which in some parts has a height of 7,000 feet, their altitude is much less imposing than it would otherwise appear. Indeed, few of the mountains lift their snowy crests more than a mile above the neighboring surface; and at intervals, in many cases of not more than 50 or 100 miles, the chain is broken by passes, affording more or less practicable routes for travel.

The Great Western Plateau presents a varied surface, consisting of elevated plains traversed by mountain ranges, with numerous broad valleys and terraced slopes.

South and east of the Great Basin it is intersected in many parts by cañons; some of them formed apparently by a violent sundering of the rocky strata whose ragged edges make up their almost perpendicular walls; others, by the slow wearing away of their beds by the streams which traverse them. Many of these cañons are of the most startling character. On some sections of the upper Colorado they are over a mile in depth, and so narrow at the top that the beams of the vertical sun scarcely irradiate the fearful chasm.

The surface of the Great Plateau also is strikingly diversified in the south by elevated tables termed mésas (*ma'sahs*), the abrupt edges of which, in many instances, are scarcely less precipitous than the rocky walls of the cañons.

The maritime chain of mountains which forms the western boundary of this plateau is in many parts higher than the Rocky Mountains. Unlike the latter, however, it rarely takes on a plateau-like form, but consists mainly of parallel ridges rising to a great altitude above their base, and including numerous snow-capped peaks. These ranges are generally sharp, and are peculiarly abrupt on the western slope. But notwithstanding their height, they are sundered at several points by transverse valleys which afford outlets to the principal rivers.

The Pacific coast region consists chiefly of narrow valleys opening toward the sea.

MINERALS.

The mineral resources of this country surpass those of any other known section of the globe. Deposits of all the principal metals except tin are found here in great abundance and richness: and those of less important metals to such extent as warrants the belief that they are not less plentifully distributed than in other countries.

The western part of the United States seems almost to realize the extravagant anticipations of the early adventurers who visited America in search of the mythical *El Dorado*. Its gold-fields are the most spacious and productive in the world, and its newly-discovered silver regions in the south also are of unusual extent and richness.

Iron is very widely disseminated. Indeed, there is scarcely a State or Territory in the national domain where it may not be mined in sufficient quantity for future wants.

NOTE.—For further information concerning the metallic products of this country, see pages 106, 107.

But the coal-fields of the United States are among its richest possessions; both on account of their unrivaled extent, and of the situation of several in the midst of vast level regions where the supplies they afford will prove of inestimable value for purposes of steam manufacture.

Those in the eastern half of the country are estimated to cover an area of not less than 225,000 square miles—nearly equal to the whole of the Western States north of the Ohio River and east of the Mississippi. The principal one extends along the western border of the Appalachian highlands, from New York to Alabama, covering nearly 100,000 square miles. Another, more than half as large, occupies south western Indiana and most of Illinois; another, of great extent, occurs in Missouri and Iowa; and another, of several thousand square miles, in Michigan.

Most of the coal derived from these fields is of a bituminous character. In some parts of the coal regions, especially the northern portion of the Appalachian field, springs of petroleum or rock-oil abound, the recent discovery and opening of which has proved of immense value in cheapening supplies for illuminating and other economical purposes. An extensive bed of anthracite coal, which affords most of the fuel for the cities of the Atlantic slope, exists in Pennsylvania, between the Blue Ridge and eastern branch of the Susquehannah.

Salt springs are common along the western border of the Appalachian highlands and in the dry regions beyond the Mississippi. In Louisiana an extensive bed of rock-salt has been recently discovered, and numerous beds of like character are believed to exist in the great table-lands of the West, especially where the systems of drainage are imperfect.

SOIL.

The soil of the United States, except of the great table-land belt west of the 100th meridian, is in general remarkably fertile. Nowhere else in the temperate zones is there a region of equal extent which rivals in productiveness the eastern half of this country. The fertility of so great an area is due not only to its favorable mineral character in most parts, but also to an ample fall of rain over its whole surface, and to its immense level tracts suited to the retention of moisture, and which in the lapse of ages have become overspread to an extraordinary depth with vegetable mold.

This fertility, however, is not without exception. Most of New England and of New York, north of the Mohawk River, has a thin and stony soil, of inferior productiveness,—being formed upon the older and harder rock—such as granite, gneiss, quartz, etc. But owing to the ruggedness of these regions, many of the valleys, and especially the low tracts along the rivers, are amply enriched with alluvial deposits washed down from the hillsides, and thus are rendered well deserving of cultivation. A great part of the surface, however, is better suited to grazing than to agriculture.

The low coastwise plain also which extends from the mouth of the Hudson River to the Rio Grande is, to a great extent, sandy and sterile; but in many sections it contains beds of marl and other fertilizing earths which may be used to render it productive. The swamp-lands, moreover, which abound along the coast, are in many parts susceptible of drainage and cultivation; and their vicinities, as well as the borders of the rivers, present alluvial tracts of superior richness.

The region of the Alleghanies is of course inferior in fertility to the level or undulating lowlands of the Mississippi valley; yet the hilly country, in most parts, abundantly rewards cultivation, while the long Appalachian valley contains some of the best farming lands in the United States.

The general character of the table-land region west of the 100th meridian—comprising the Western Steppes and Great Western Plateau—is that of marked sterility. This is due mainly to the lack of rain, especially during summer; but in some parts the natural constitution of the soil is also unfavorable to productiveness.

Near the mountains, however, and in the bottom-lands along the rivers, there are many tracts suited to cultivation, while the neighboring sections generally support a plentiful herbage and are well adapted to grazing. But more remote from these regions, except along the eastern border, the plains are strikingly barren; and the hollows in some parts are covered with a saline or alkaline crust, caused by want of drainage and by the evaporation of water which collects in them from the neighboring slopes, and thus is impregnated with the soluble matter of their soil. Nearer the mountains, where the supply of water is somewhat more abundant, there are many salt pools. The valleys known as *Parks*, between the parallel ranges of the Rocky Mountains, are remarkable for their beauty and fertility.

The Pacific coast-region is distinguished for the highly fertile intervals between its mountain spurs, which, owing to the peculiar favorableness of the climate, are almost unrivaled in their capacity for producing the smaller cereals, root-crops, and fruit. South of Monterey Bay, however, the country is somewhat arid; and in many parts, especially where facilities for irrigation are scanty, is better suited to grazing than tillage.

VEGETATION.

Owing to the great extent of its well-watered and fertile plains, this country is peculiarly rich in its vegetation.

It is especially distinguished for the extent and variety of its forests. Nearly the whole eastern section, from the Mississippi to the Atlantic, is, in its native state, richly wooded.

The western or Pacific slope, on the other hand, from the northern boundary to about the 40th parallel, except in the more arid districts, is also clothed with fine forests, and still farther south presents luxuriant wooded tracts along the maritime border.

A marked contrast, however, appears between the forests of the east and west. In the north, an extensive belt, chiefly of white pine, of great economic value, reaches from the Mississippi valley to the Atlantic, its southern limit being in about the latitude of the northern shore of Lake Ontario.

South of this belt, as far as to a little below the mouth of the Ohio River and Chesapeake Bay, the forests are mainly of deciduous trees. These forests are remarkably mixed; comprising maple, beech, oak, chestnut, hickory, ash, elm, birch, cherry, bass-wood, buckeye, etc., etc. In the richness of their foliage, as variegated by autumnal tints, they are probably unrivaled in any section of the globe.

Entering the warmer latitudes below the last-mentioned limit,* foliaceous or broad-leaved evergreens become common, and vegetation in general assumes a sub-tropical

* This limit nearly coincides with the northern boundaries of Tennessee and North Carolina.

character. Live-oaks appear along the Atlantic coast; cypress and gum trees grow numerous; and soon the characteristic region of magnolias and laurel is reached.

The dry lowlands of the Atlantic and Gulf slopes are overspread, to a great extent, with the long-leafed pitch-pine, one of the most picturesque and valuable timber trees of this section.

Among other characteristic forms are the southern cotton-wood, catalpa, bumelias, palmetto, persimmon, and Chickasaw plum. Various species of oak, maple, beech, etc., abound in the uplands and elsewhere.

In the Peninsula of Florida, vegetation is more tropical. Mangroves appear in extensive thickets; several characteristic fruits, among which is the Florida orange, are introduced; and various straggling forms from the neighboring zone become more or less prevalent.

It is evident from the foregoing that the forests of the eastern part of the United States are remarkable not only for their extent, but also for their variety. In fact, they have "no parallel for the diversity of species collected in a growth of trunks of nearly the same size, and thriving on the same soil and in the same climate."

The western slope of the country presents a vegetation strikingly dissimilar to that of the eastern. On the Pacific side, deciduous trees are decidedly repelled, and the number of their species is comparatively few.

Of the leading forms prevalent in the eastern forests, the following are entirely absent in the western; viz., magnolias, bass-wood, locust, and other pod-bearing trees, elms, walnuts, hickories, beeches, etc. The chestnut is represented merely by a single species, and the maples by only one or two small forms.

Cone-bearing trees, on the contrary, except in the arid districts, are particularly favored, "affording the finest evergreen forests known in the temperate latitudes. Many of these trees, in the moist and equable climates near the Pacific, are of gigantic size; their height and vastness of trunk striking the beholder with astonishment.

The pine family, especially, are remarkable in this respect. Among these, the Douglas spruce, the sugar-pine, a species of yellow pine, and the balsam-fir tower to a height of over 200 feet. A species of white cedar, also, has a similar size.

But the majestic redwood family surpass all others. The common redwood grows to an altitude of from 180 to 300 feet, often with a thickness of 10 or 12 feet; while the giant redwood (confined to a single small locality in California) reaches the wonderful height of from 300 to 400 feet and a diameter of from 20 to 30 feet. These trees have no rivals in the existing vegetable kingdom.

Among the latter species, there is a single prostrate trunk (known as the Monarch of the Forest), the top of which has been partly destroyed—probably by fire—which measures 36 feet in diameter at the butt, and 110 feet in circumference. The length of the whole portions of the trunk is 300 feet, and the diameter at the small end 12 feet. From fragments in a line with the main stem, and apparently belonging to it, it is estimated that its full height when standing must have been from 450 to 500 feet.

Turning our attention to a different field, it may be observed that notwithstanding the vast extent of forests in this country, there are immense areas nearly or quite destitute of trees. Of this character are the dry regions of the Great Western Plateau and nearly the whole of the Steppes, except the bottom-lands along the rivers. Of the same description, also, is a large part of the Mississippi lowlands, extending from the Steppes to the northwest corner of the Gulf of Mexico, and farther north, invading the central portion of the Mississippi valley nearly as far as the middle course of the Ohio River.

These treeless regions are called prairies. They are covered, for the most part, with coarse grass, intermixed on the table-lands—in the warm-temperate and cool regions—with deep-rooted shrubs, as the artemisia (inappropriately called sage*), and—in the subtropical and tropical regions—with similar shrubs and thick-leafed juicy plants, such as the cactus and yucca (or Spanish bayonet).

The absence of trees on the table-lands is doubtless owing chiefly to the dryness of the climate and the consequent aridity of the soil.

In the less elevated regions of the Mississippi valley it may be partly due to the same cause and partly to the extraordinary depth of the soil, which in many places probably renders the proportion of moisture retained near the surface insufficient for the support of a forest growth. But the treeless character of these plains is believed to be mainly owing to their visitation in former years by sweeping fires, which are peculiarly liable to originate on the dry grassy table-lands, and spread eastward as far as local conditions permit.

The United States possesses a numerous assemblage of interesting grasses. Among the most remarkable is the tree-like cane, which forms extensive brakes or thickets along the Gulf of Mexico and in the moist alluvial lands on the borders of the rivers and elsewhere throughout the South. So dense are these brakes in many sections that they are absolutely impenetrable except with the aid of the hatchet.

The adaptation of the United States to the production of different agricultural staples varies to a remarkable degree with its soil and climate. In respect to several staples, it holds a rank attained by no other country. Its capacity for yielding breadstuffs in the North and cotton in the South is unrivaled, while its suitableness to the growth of tobacco in the middle latitudes probably surpasses that of any other region of equal area.

The more hardy cereals (maize, wheat, oats, rye, and barley) flourish throughout the agricultural sections. Maize, however, is the most characteristic staple, and by far the most important. It grows most luxuriantly in the Southern and Western States, and is of especial importance in the central portion of the Mississippi valley. In the maritime region of the Pacific, however, it thrives less successfully, and in many localities fails to mature, owing to the prolonged drought and cool nights of summer.

Wheat and oats, which rank second and third in importance among cultivated cereals, grow best in the valleys along the Pacific, in the Western States north of the Missouri and Ohio rivers, and in the Middle States and those of the Southern States adjoining the latter. The yield of these grains per acre, however, is most abundant in the two former sections.

Rice finds a congenial locality in the marshy tracts along the southern seaboard, particularly in South Carolina and Georgia.

Cotton flourishes chiefly south of the 35th parallel, in the States north of the Gulf of Mexico, or bordering the Atlantic.

Tobacco has its most profitable range between the 36th and 40th parallels. [*See this range on the map.*]

The potato is a characteristic product of the northern portion of the country, particularly of New England and the Middle States; also of the Pacific slope west of the Cascade Range and Sierra Nevada.

The batatas or sweet potato thrives in the warmer regions.

Sugar-cane appears to have found only a few limited sections where it is a preferable crop, its culture thus far having been confined chiefly to that part of Louisiana projecting south of Mississippi.

The portions of the United States well suited to the production of hay-crops or to grazing are of immense area. Throughout the fertile districts, except in the warmer parts of the Southern States, the unplowed fields are overspread with a rich carpet of verdure which affords excellent pasturage during the warm months, or yields a valuable harvest for winter store. The profitableness of the latter may be inferred from its ranking third in value among the crops of the country, being surpassed in this respect only by corn and cotton.

The dry table-lands, also, except in the most arid districts, afford good grazing during a large part of the year; and in many localities where the grass is somewhat abundant, the summer drought causes it to cure or become hay while standing in the field; so that ample pasturage is afforded (the snow being light) through all or nearly all winter.

ANIMALS.

The northern border of the United States is embraced in the district of fur-bearing animals, and the forest tracts of the northeast and northwest contain various species which are much sought by the hunter and trapper. The most important on account of their furs are the otter, beaver, mink or American sable, marten, and musk-rat. There are also found in the same range the black bear, wolverine, Canada lynx, bay lynx or wild-cat, and wolf. Several of the above, however, as the Canada otter, black bear, and wild-cat, likewise dwell far south. The moose-deer exists in the extreme northeast.

The eastern forest regions are inhabited by the common American deer, the American elk or stag (now very rare in this section), the Virginia opossum, raccoon, etc. The common American deer likewise frequents the grassy regions of the Great Central Plain.

But in general the prairies have a fauna somewhat peculiar. Their most noted habitant is the American bison or buffalo, which roams in vast herds over the unfrequented tracts west of the Mississippi. The prairie-wolf, also, is especially characteristic of these plains. Upon the plateaus and within the Rocky Mountain district the American elk is again found; and in the same range the mule-deer, prong-buck, American badger, etc.

The mountain sides are inhabited by the Rocky Mountain goat and big-horned sheep, the latter roving throughout the rugged highlands from the Western Steppes to the Pacific.

The species already mentioned as belonging to the prairies are found both east and west of the Rocky Mountains. It must be remembered, however, that, as a whole, the faunas on opposite sides of this chain are different. Thus part of the ruminants, the gnawers, the insects, and all the mollusks are of distinct species.

From the foregoing it will be observed that most of the principal ruminants are found west of the Mississippi, where the unfrequented grassy plains afford them abundant subsistence. The range of several, however, formerly extended much farther east—the buffalo, for instance, having been known even upon some parts of the Atlantic coast.

The largest and most powerful of American carnivora—the grizzly bear—has a similar range with the above, preying upon those whom feebleness or want of fleetness renders unable to escape. His chief home, however, is the oak and pine regions of the Western slope, where he finds an agreeable subsistence of acorns and pine-cones as well as flesh. The puma or cougar (known also as the panther or American lion, but unlike either the lion or panther of the Old World) frequents the same range as the grizzly bear; and, to a greater or less extent, wanders through nearly all the unsettled parts of the country.

The United States is frequented by a great number and variety of migratory birds, the principal of which are wild geese, ducks, and pigeons. The wild turkey is common in the lowlands of the Mississippi valley.

The principal birds of prey are various species of eagles, hawks, owls, and buzzards. The turkey-buzzard, in particular, is numerous in the southeast, where it frequents even the populous settlements, and serves as a valuable scavenger.

The most noted reptiles are turtles, alligators, and rattlesnakes. Alligators infest the rivers, marshes, etc., of the warm regions in the east and southeast; their range extending as far north on the coast of South Carolina as latitude 32° 30′.

Here, as elsewhere, noxious insects abound in the heated districts. Mosquitoes, especially, are an annoying pest, being so numerous about the rivers and marshes in some parts of the south as to render the vicinity uninhabitable.

The coasts of this country are well supplied with fish. Cod, haddock, mackerel, herring, halibut, etc., are abundant. The inland waters are likewise well stocked; white-fish abound in the Great Lakes; and pike, trout, perch, etc., are common in both the large and small bodies of water.

CONCLUSIONS.

From the foregoing account of the physical geography of the United States, it is obvious that this country possesses within itself the natural resources for an extraordinary material prosperity. The remarkable fertility of its soil; the unequaled richness of its mineral deposits; the ample facilities for manufactures afforded by the numerous rapid streams which descend from the uplands along its coasts, or by the inexhaustible supplies of coal distributed through its interior; and the extraordinary advantages for commerce presented not only by its extensive coast line and its vast network of inland waters, but also by its broad plains easily crossed by railways or trenched by canals—all combine to render it of surpassing fitness for the abode of man.

Hence, peopled in the main by a highly cultivated race, the varieties of which are fast becoming intermixed here, and thus are producing an unusually vigorous stock, it remains only to develop these resources under a just and wise policy in order to attain a national greatness unparalleled in history.

* Artemisia-tridentata. Many parts of the table-lands, overspread with this plant, are termed "sage plains."

APPENDIX II.

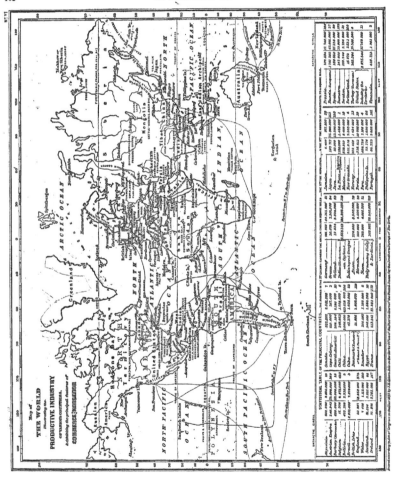

QUESTIONS ON THE MAP.

Note.—As many branches of industry consist in a mere development of the natural resources or capabilities of the regions where they are prosecuted, they show forth, in a striking manner, the physical relations of these regions to man. A reference, therefore, to the principal seats of such branches of industry, as exhibited on the map, will be found useful in the present connection. The following questions, although necessarily incomplete, will suffice to show how the investigation of the subject may be pursued.

What fishery is the leading branch of industry in the high northern latitudes ? What valuable fossil remains are obtained in the northern part of Siberia ? What does the presence of this fossil together with accompanying relics show ? *Ans. That northern Siberia, at an earlier geological period, was inhabited by a species of elephant which has since become extinct.* The obtaining of what vegetable products near the Arctic Circle, in Norway and Sweden, shows that Western Europe has a mild climate in the high latitudes ? What rich materials for warm clothing are obtained chiefly in the cold regions of North America and Asia ? Why are the furs from these regions preferable to those from warmer districts ? *Ans. Because they are thicker and finer; nature having provided a superior covering to protect the animals of these regions from the cold.*

What metal, in the New World, is obtained very abundantly from Chile and the vicinity of Lake Superior? What, principally, from the western part of the United States? What precious stone from Brazil? For its yield of what minerals is Great Britain especially distinguished? For what the Ural mountain-region, between Russia and Siberia? For what precious metal is Australia famous ?

What material for the manufacture of dress-goods, etc., is most abundantly produced in the southern part of the United States? What rich material for the same purpose is largely produced in the Mediterranean countries of Europe? What other fibrous material is produced in Spain, of the finest quality? For the yield of what fibrous materials is Russia especially noted? In what parts of Asia is silk an important product? In what portion of Africa is wool a noted staple? In what great island is its yield likewise of much importance? In what portion of South America is it a chief staple? What is the general character of the great wool-growing regions? *Ans. They are regions of a dry climate, chiefly suited to pastoral employments.* What else is generally a principal product of the same regions? *Ans. Hides.*

In what country of South America is coffee produced more abundantly than elsewhere in the world? Of what part of the world is it an indigenous product? *Ans. Of the uplands of Eastern Africa.* In what archipelago are spices most abundantly obtained? In what peninsula of southern Asia is opium chiefly produced?

In what desert does rock-salt abound? How does the dryness of the desert favor the formation of salt beds? *Ans. By evaporating the moisture that collects in the hollows, and which brings thither in solution saline matter gathered from the neighboring soil, leaving it behind as the aqueous particles pass off into the atmosphere.*

CHIEF PRODUCTIONS OF THE EARTH, AND THE COUNTRIES WHERE THEY ARE PRINCIPALLY PRODUCED.

Almonds.....Syria, Tripoli, Barbary, Spain, Portugal.
Aloes.........Socotra, Arabia, Barbadoes, South Africa.
Allspice.......Jamaica.
Amethysts ....Brazil, Siberia, Ceylon.
Arrow Root....South America, East and West Indies, South Sea Islands.
Barley........Between latitude 69½° and 45° Eastern Hemisphere, British America, and Australia.
Bread Fruit...Polynesia, East Indies.
Cassia.........East and West Indies.
Chestnuts.....Spain, Italy, Corsica, Turkey.
Cinnamon.....Ceylon, Cochin China.
Citron........Madeira, Polynesia.
Cloves........Molucca Islands.
Coal..........Britain, Belgium, United States, Australia.
Cochineal.....Mexico, Central America, West India Islands.
Cocoa.........West Indies, South America.
Cocoa Nuts ..Ceylon, Maldive Islands, Siam, Bengal, Brazil, Polynesia, Africa.
Coffee........Arabia, Java, West Indies, Brazil, Mauritius.
Copper........Britain, Chile, United States, Sweden, Siberia, Persia, Japan.
Cork..........France, Spain, Portugal, Italy, Barbary.
Cotton........Grows naturally in Asia, Africa, and America ; it is much cultivated in the warmer parts of the United States and elsewhere.
Currants......Ionian Islands and Greece yield the small dried grapes commonly called currants.
Dates.........Egypt, Barbary, Arabia, Persia.
Diamonds.....Brazil, Borneo, India.
Ebony.........Mauritius, Madagascar, Ceylon.
Emerald ......Peru.
Figs...........Turkey, Greece, France, Spain, Italy, North Africa.
Flax..........Russia, Egypt, Ireland, Netherlands, New South Wales.
Furs..........British and Russian America, Russia, United States.
Gamboge......Siam, Cambodia.
Gold..........United States, Australia, India, Russia, Africa, Hungary, Saxony, Equador.
Hemp.........Russia, Italy, Philippine Islands, Brazil, Britain, Egypt, North America.
Indigo........East and West Indies, Guinea.
Ipecacuhana...Brazil, South America.
Iron..........Most countries, particularly Britain and the United States.
Ivory........Africa, East Indies.
Lead..........Britain, United States, Germany, Spain.
Lemons ......Syria, Persia, Greece, Italy, Spain, Portugal, Azores, West Indies.
Mace ,........East and West Indies.
Mahogany.....West Indies, Central America.
Maize, or Indian Corn. } America, from Canada to La Plata, South Europe, Central Africa, Australia.
Maple Sugar...Canada, United States.
Marble........Italy, Greece, Egypt, Britain, France, United States.
Mercury ......Spain, Austria, California, Peru, China.
Millet.........Germany, Poland, India, Africa.
Molasses......West Indies, Mauritius, Louisiana.
Morocco ......Levant, Barbary, Spain, Flanders.
Mulberry and Silk Worm as } South Europe, South Asia.
Nutmeg.......Moluccas, Sumatra, Penang, Borneo.
Oats.........Chiefly grown in latitudes north of Paris, though cultivated in Bengal as low as the 25th degree.
Olives........Syria, Greece, Africa, Spain, Italy, Brazil, Ionian Islands.
Opals.........Hungary, East Indies.
Oranges:......The Azores, Spain, Portugal, China, Italy, Malta, Polynesia, West Indies.
Palm Oil......Western Africa, Fernando Po, Brazil, Hindoostan.
Pepper........East and West India Islands, French Guayana.
Pine Apples...West Indies, Mauritius, Hindoostan, Polynesia.
Plantains, or Bananas... } Tropical America (especially in Mexico), Polynesia, East Indies.
Platina........Spain, Asiatic Russia, South America.
Pomegranates .Persia, South Europe, Tropical Asia, West Indies.

Prunes ......South France.
Raisins ....Smyrna, Valencia, Malaga, Italy.
Rice..........India, China, West Indies, United States, Italy, Africa.
Ruby........Ava, South America, Siberia, Egypt.
Rye ....The Bread-corn of Germany and Russia.
Sago.........East Indies.
Sapphire......Ava, Bohemia, Saxony, France.
Silver ....Mexico, Peru, United States, Hungary, Saxony, Siberia.
Sponge ....Found upon the rocks of the Mediterranean and Red Sea.
Sugar Cane....Tropical America, East and West Indies, Louisiana, Sicily, Canary Islands, Polynesia, Africa.
Sugar from } France, Belgium, Germany, Prussia, Russia.
Beet Root. } F
Tamarinds ...East and West Indies, Arabia, Egypt, Cuba, Brazil.
Tapioca......South America, West Indies.
Tea...........China, Japan, Assam.
Tin...........Cornwall, Devon, Galicia, Erz-gebirge Mountains in Saxony, Bohemia, Malay, China, Island of Banca in East Indies.
Tobacco.......Tropical America, United States, Turkey, Asia, Prussia, France, Australia.
Topaz........South America, India, Egypt, Siberia, Mexico.
Turquoise.....Nishapore in Persia.
Vine.........South Europe, Canary Islands, Africa, North America in latitude 46°, Brazil.
Wheat........We are in total ignorance where this important grain was first cultivated; some suppose in Northern Africa. It is raised in almost every part of the temperate zones. Little is grown beyond latitude 58° in Europe, but on the Alps it ripens to the height of 3,500 feet above the level of the sea.
Wines.........Port.—Province of Upper Douro, in Portugal. Sherry.— Xeres, near Cadiz, in Spain. Claret.—Bordeaux, in France. Champagne.—From a province in France of the same name. Burgundy.—Ditto. Madeira.—From the Madeira Islands. Malmsey.—Ditto. Teneriffe.—From the island of Teneriffe. Marsala.—Sicily. Cape.—From South Africa.
Yams.........Africa, South America, Polynesia, Australia.

EXPORTS OF COUNTRIES.

EUROPE.

Russia...........Timber, deals, tallow, corn, hemp, flax, furs, linseed, hides, leather, pitch, tar, wax, feathers, pearl-ashes.
Spain and Nor- } Timber, deals, iron, pitch, tar, turpentine, resin, oak
way.........} bark, juniper-berries, and fish.
Germany........Wheat in large quantities from Dantzic; hemp, flax, wool, bark, amber, Rhenish wines, hops, toys, etc.
Denmark.......Hogs, rape-seed, fish, and feathers.
Holland and } Butter, cheese, spirits, flower roots, madder, hops, lace
Belgium.....} and linen, clocks, toys, etc.
France..........Wines, brandy, fruits, silks and gloves, perfumery, trinkets, and fancy articles.
Spain...........Wine, fruits, olive oil, cork, wool.
Portugal........Wine, fruits, cork.
Italy ....Raw and manufactured silks, fruits, olive oil, straw-plait, cheese, maccaroni, vermicelli, sulphur, pumice stone, marble, paper rags.
Greece..........Raw silk, dried fruits.
Turkey ....Leather, raw silks, figs.

ASIA.

Hindoostan......Silk, opium, sugar, coffee, pepper, indigo, rice, lac-dye, saltpeter, precious stones.
Birman Empire..Teak timber, rice, indigo, gums, drugs, palm sugar, cotton goods, silk, varnish.
China,..........Tea, silk, cotton goods, porcelain, lacquered ware, gums, paper, drugs.
Japan ....Silks and cotton goods, drugs, spices, varnish, porcelain, rice, cedar.

Persia ....Silks, carpets, cotton goods, shawls, sugar, rice, dried fruits, leather, drugs, tobacco.
Arabia............Coffee, aloes, gums, myrrh, frankincense, perfumes, drugs.
Turkey in Asia.....Coffee, carpets, silks, fruits, drugs, opium.
Siberia............Metals, precious stones, leather, and furs.
Kamtchatka ......Furs and dried fish.
Asiatic islands.....Cinnamon, cloves, nutmegs, pepper, ginger, sago, camphor.

AFRICA.

Mauritius.........Palm oil, teak timber, aloes, dye-woods, ostrich feathers, ivory, gold, sugar.
Morocco..........Leather, goat-skins, gums, fruits.
Algiers and Tripoli.Ostrich feathers, dates, wax, wool.
Egypt ....Cotton, indigo, drugs, fruits, wheat, rice.
Madeira Islands....Wine, fruits.
Canary Islands.....Wine, fruits, silks, barilla.

NORTH AMERICA.

Canada ....Timber, wheat, pot and pearl ashes, furs, fish.
Newfoundland.....Cod and other fish.
Nova Scotia ......Timber, dried fish, plaster of Paris.
Hud. Bay Territory..Furs.
United States :
Eastern States...Cotton and woolen goods, boots and shoes, metallic wares, lumber, beef, fish, oil.
Middle States....Grain, flour, dairy produce, iron merchandise, clothing, coal, rock oil.
Southern States ..Cotton, tobacco, lumber, naval stores, rice, sugar.
United States :
Western States...Grain, flour, beef, pork, lumber, copper, lead.
Territories and } Gold, silver, timber, furs.
Pacific States }
Mexico............Silver, cochineal, vanilla, sarsaparilla, hides.

SOUTH AMERICA.

Caraccas..........Cocoa, coffee, indigo, tobacco.
Guayana ....Sugar, rum, cotton, coffee, tobacco, indigo, cayenne pepper.
Brazil..............Cotton, sugar, coffee, tobacco, dye-woods; drugs from the northern provinces; gold and diamonds from the middle; and wheat, hides, and tallow from the southern.
Buenos Ayres......Gold and silver, hides, beef, and tallow.
Peru...............Silver, gold, alpacca hair, cinchona, hides, guano.
Chile...............Silver, gold, and copper from the northern provinces; wheat and hemp from the southern.
West Indies ....Sugar, coffee, rum, molasses, cotton, pimento, ginger, logwood, mahogany, cocoa, cochineal, cigars.

TRADE ROUTES.

The navigation of the ocean constitutes an important branch of industry, in which a greater or less number of people of all civilized countries are engaged. The most wealthy and powerful nations are those which have the most extended foreign commerce; as, Great Britain, France, the United States, Holland, Denmark, Sweden, and Russia. Commerce has always been a fruitful source of individual and national prosperity.

In former times, maritime pursuits were very slowly conducted. This was owing to the imperfect construction of vessels (which were built more with reference to strength than qualities of fast sailing), and to the prevailing ignorance of the winds and currents of the ocean, and how the mariner might best avail himself of them in steering his vessel from one part of the world to another.

The winds and currents of the ocean have formed subjects of the most careful study and research, the results of which have been of the greatest utility to all engaged in navigating the sea. No seaman is qualified to direct the course of a ship who does not know where prevailing winds and currents are to be met with, and how to turn them to the best advantage in prosecuting his voyage.

On this subject Captain Basil Hall remarks : " It is one of the chief points of a seaman's duty to know where to find a fair wind, and where to fall in with a favorable current. If we take a globe and trace on it the shortest route by sea to India, and then fancy that such must be the best course to follow, we shall be very much mistaken. And yet this is very much what our ancestors actually did, till time and repeated trials, and multitudinous failures, gradually taught them where to seek for winds, and how to profit by them when found."

Map 6 exhibits the tracks usually taken by ships proceeding from New York across the Atlantic, Pacific, and Indian oceans. The *outward* and *homeward* tracks are distinguished by arrows.

ROUTE FROM NEW YORK TO SAN FRANCISCO.—The route marked on the map shows the course taken by vessels which have made the quickest passages between these ports. The pupil will notice that it is not the shortest as regards distance. Between New York and the point where it crosses the equator (on or near the 30th meridian), and also between Cape Horn and San Francisco, it diverges very considerably from a direct line.

To understand this route, and others marked on the map, it is necessary for the learner to bear in mind the direction of the prevailing winds as explained in Lesson IV., Part III. From the parallel of about 30° north and south, nearly to the equator, there are two zones of perpetual winds, namely : the zone of northeast trade-winds on this side, and of southeast trade-winds on that. Now, a vessel sailing from New York to Cape Horn is necessarily obliged to pass through these zones. Before striking the northeasterly trades, she must make a good deal of easting, that is, proceed to the east ; for if this is not done they would, perhaps, carry the vessel too close to the Windward Islands and the northern coast of South America, so that she would find it very difficult to double Cape St. Roque.

After crossing the equator, the route extends through the South Atlantic, at no great distance from the eastern coast of South America, passing inside the Falkland Islands. The most difficult part of the route is that which extends from the 50th parallel in the South Atlantic to the same parallel in the Pacific. In this part of the voyage is performed the labor of doubling Cape Horn, a very troublesome operation in consequence of the continuous cold westerly winds which sailors always find there. The best months for doubling the Horn are our winter and summer, excepting July. October appears to be the most unpropitious month for the passage.

After reaching the 50th parallel, the California vessel stands far out into the Pacific at a great distance from the coast. This is done to get the southeast trades in their full force, for, it is to be remembered, these winds are considerably impeded by the continent, and are the strongest and steadiest at a distance from shore. On passing the region of calms, near the equator, the zone of the northeast trades is met with, and in crossing this belt, the track, instead of leading directly to San Francisco, continues on still in a northwest direction until the vessel, in about the 35th parallel north latitude, has got beyond the influence of these trades ; then easting is made and the port reached.○

Returning from San Francisco, a vessel pursues a course nearly due south, between the meridians of 120° and 125° west longitude, to about the 50th parallel south latitude, where westerly winds are met, which rapidly bear her past Cape Horn into the South Atlantic. Here her course is rather slow and irregular until Cape St. Roque is reached. At the latter point she enters the strong current which sweeps westwardly from the Gulf of Guinea, and flows along the northern coast of South America. (Art. 364, 365.) From the equator to New York the track is very nearly straight.

FROM NEW YORK TO PORT PHILIP, AUSTRALIA.—The track of vessels bound from New York to Australia is the same as that pursued by ships bound for California, until the 20th parallel south latitude is reached. Indeed, all vessels sailing for the South Atlantic, whether their destination be Rio Janeiro, San Francisco, Cape of Good Hope, or Port Philip, are obliged to follow about the same course until they have passed the latitude of Cape St. Roque.

The following remarks relating to this route are taken from the " *Sailing Directions ;*" " The gold ports of Australia, whether the distance be measured *via* Cape Horn, or by the way of the Cape of Good Hope, are between

* The track above described is sometimes departed from, more or less, by vessels bound from New York to San Francisco ; but it is the one recommended in the " *Sailing Directions.*" It is very nearly the course taken by the ship " Flying Cloud" on the trip she made in 90 days—the quickest passage ever performed between these ports.

12,000 and 13,000 miles from the Atlantic ports of the United States or Europe. The best way for vessels in the Australian trade, from Europe or America, *via* the Atlantic, to *go*, is by doubling the Cape of Good Hope ; and the best way to *come* is, *via* Cape Horn ; and for this reason, viz. : The prevailing winds in the extra-tropical regions of the southern hemisphere are from the northwest, which of course makes fair winds for the outward bound around the Cape of Good Hope, and fair winds for the homeward bound around Cape Horn. Here, all is plain sailing ; vessels homeward bound should steer by the shortest cut for Cape Horn, and the outward bound, after doubling the Cape of Good Hope, should shape their course as direct for the port of destination as the land and winds will permit them."○

Many of the other routes, as marked on the map, appear to be very circuitous, and some of them are actually so ; but they are such as the long experience of seamen have found to be the best, and such, too, as would be pursued, without much experience by a commander of a vessel who was fully acquainted with the regular movements of the air and ocean. The limits of this book do not admit of a further explanation of the tracks of vessels. By perusing Lesson XVI., Part II., on the Currents of the Ocean, and also Lesson IV., Part III., on the Permanent Winds, the learner may be able to understand why particular deviations from a direct line are made in the several routes marked on the map.

METALLIC PRODUCTIONS.

Of the great number of metallic substances found in the earth, the most useful are gold, silver, mercury, tin, copper, zinc, lead, and iron.†

Metals are deposited in veins or fissures of rocks, in masses, in beds, and sometimes in gravel and sand. Most of the metals are found in veins ; a few, as gold and tin, iron and copper, are disseminated through the rocks, though rarely. The veins are cracks or fissures in rocks, seldom in a straight line, yet they maintain a general direction, and sometimes extend to an unfathomable depth.

Metals are peculiar to particular rocks : gold and tin are most plentiful in granite and the rocks lying immediately above it ; copper is deposited in various slate formations ; lead is found in the mountain-limestone system ;

SHAFT.

iron abounds in the coal strata ; and silver occurs in almost all these formations ; its ores being frequently combined with those of other metals, especially of lead and copper.

When a mine is opened, a shaft like a well is sunk perpendicularly from the surface of the ground, and from it horizontal galleries are dug at different levels according to the direction of the metallic veins. When mines extend very far in a horizontal direction, it becomes necessary to sink more shafts, which are connected together by horizontal galleries. Shafts are from eight to twelve feet square, and are usually walled up with timber or stone to prevent the sides from caving in. The water which filtrates through the earth would soon collect into a mine and put a stop to the work, were

* This work has been prepared for several years under the direction of the Superintendent of the National Observatory at Washington. It is designed to accompany and explain the " *Wind and Current Charts,*" issued from the same source. Both have been published at the expense of the United States Government, and distributed gratuitously to the commanders of all vessels who have pledged themselves to keep a journal of their voyages, and, on their return, to transmit the same to the National Observatory.

† Thirty-five metals are now known : they are gold, silver, platinum, copper, lead, tin, iron, zinc, arsenic, bismuth, antimony, nickel, quicksilver, manganese, cadmium, cerium, cobalt, iridium, uranium, chrome, lantanium, molybdenum, columbium, osmium, palladium, pelapium, tantalum, tellurium, rhodium, titanium, vanadium, tungsten, dydynium, ferbium, orbium.

not adequate means employed· to remove it. This is done in two .ways ;
first, by digging a horizontal gallery (called an *adit-level*) from the mine to

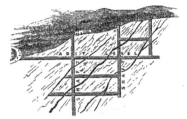

MODE OF OPENING A MINE.

some adjacent valley ; in this way is drained all that part of the work
which lies above ; and, secondly, by the use of pumps for drawing out the
water·from that part of the mine which is below the adit-level, or drain.
Many of these drains are of great length, and are dug at great expense.
One.of these, in the mining region of Cornwall, England, begins in a valley.
near the sea, and a very little above its level, and goes through all the
neighboring mines, which it drains to that depth, and with all its ramifica-
tions is thirty miles long.

The mode of raising ores varies with the depth of the mine. For a distance of fifty feet, or even one hundred, the ore and rubbish may be raised
to the surface by the simple windlass, worked by hand, on which a rope is
so wound that one bucket descends while the other ascends. As soon, however, as the depth of the shaft becomes more considerable, it is necessary to
resort to horse or steam power for raising the ore. The common machine

MODE OF HOISTING ORES.

used.for this purpose is called a whim ; as usually constructed when worked
by horse power, it is represented in the annexed ·cut. A steam-whim is
generally used when the shaft has a depth of more than two hundred feet.

NOTE.—The author is indebted for the following facts, relating to·the distribution of
metals, chiefly to the able work of J. D. Whitney, Esq., entitled, " METALLIC WEALTH OF
THE UNITED STATES."

DISTRIBUTION OF GOLD.

IN THE UNITED STATES.—The United States contains three ·principal gold-
fields ; viz., the Appalachian gold-field, the Rocky Mountain gold·field, and
the Sierra Nevada gold-field ; but they are of very unequal· importance.
That of the Atlantic slope, the " Appalachian gold-field," has been worked
to a moderate extent for about forty years ; the others are of comparatively
recent discovery, but have yielded vastly more than the former.

The Appalachian gold-field extends from Georgia, in a northeasterly direction to Maryland, and is developed in the following counties : in *Georgia*,
in Carroll, Cobb, Cherokee, Lumpkin, and ·Habersham counties ; in *South
Carolina*, through the whole northwestern corner of the State, especially in
the following districts : Abbeville, Pickens, Spartanburg, Union, York,
Lancaster ; in *North Carolina*, in Mecklenburg, Rutherford, Cabarras, Rowan,
Davidson, Guildford, and· Rockingham ; thence through *Virginia*, in Pitt-
sylvania, Campbell, Buckingham, Fluvanna, Louisa, Spottsylvania, Orange,
Culpepper, Fauquier ; in *Maryland*, Montgomery County. Gold has also
been found· in Canada on ·the Du Loup and Chaudiere rivers.

The Rocky Mountain gold-field is known to be of extraordinary richness ;

although as yet it is but imperfectly explored. The first noteworthy dis-
coveries of gold within its limits were made in 1858, in the vicinity of
Pike's Peak : since which, extensive auriferous regions have been developed on each side of the chain, both to the north and south—thus
calling into political existence the new Territories of Colorado, Idaho, and
Montana.

The Sierra Nevada gold-field is the most productive in the United States,
and, compared with what we know of other similar regions, it is perhaps
the richest in the world. It was first discovered in 1848, and it is calculated that,.up to the end of 1863, it had yielded the enormous sum of nearly
eight hundred million dollars. and that the annual product is now about
fifty million dollars. This rich mineral region is in the great valley of California, which has a length of about 500 miles, and a breadth of from 50 to
100. It is.drained by two principal rivers,—the Sacramento and the San
Joaquin, the former flowing south, and the latter north. The gold is most
commonly obtained in fine particles, and scales or flattened grains, and is obtained by *washings*, or the separation of the metal from the earthy substances
with which it is mixed. It is also procured from veins of quartz by crushing the solid rock. Lumps or "nuggets" are not common, and rarely exceed a few pounds in weight.

IN FOREIGN COUNTRIES.—Of the foreign countries, those most productive
of gold are Australia, the Russian Empire, South Asia, the East Indies,
Equador, New Guinea, Mexico, Brazil, and the Austrian Empire.

The Australian gold district is, next to that of California, by far the most
productive of any known. The existence of gold there was first made
known in May, 1851. The gold region embraces the southeast corner
of Australia. The yield, for several years, exceeded that in the United
States, but of late has ·considerably diminished. The gold of the *Russian
Empire* is obtained almost entirely from the eastern slope of the Ural Mountains, from the vicinity of the Altai Mountains, and from the Caucasus. It
is supposed that the countries of Russia yield gold to the value of about
thirty million, dollars annually.

DISTRIBUTION OF SILVER.

IN THE UNITED STATES.—Extensive silver mines have lately been opened
in Nevada·and Arizona, and are yielding large returns. The silver furnished by this country formerly came mostly from the native gold of California, with which silver is alloyed. From this source a considerable amount
is still obtained. .

IN OTHER COUNTRIES.—The countries particularly distinguished for their
extensive yield of silver are Mexico, Peru, Chile, Bolivia, and Spain. Mexico is by far the richest in mines of this metal, from which, it is estimated,
about thirty-five million dollars worth are now annually obtained—an
amount far greater than the yield of any other mines in the world.

The province of Guanaxuato is supposed to furnish about one half the
amount of silver produced in Mexico. In *Peru*, the richest mines are at Pasco,
on the Andes, over 13,000 feet above the sea. Besides the Pasco mines,
which are the richest in the world, there are numerous other mining districts in Peru, .especially in the province of Pataz, Huamanchuco, Caxa-
marca, and Hualgayoc. The richest silver mines of *Chile* are those of Copiapo ; those of *Bolivia* are at Potosi, more than 16,000 feet above the
sea-level. Of the silver furnished by Europe, that of *Spain, Russia*, and
Great Britain is derived almost exclusively from the working of silver-lead
ores. The only mining districts of importance in Europe, in which silver
ores are worked by themselves, are those of *Hungary* and *Transylvania*, of the
Erzgebirge in *Saxony* and *Bohemia*, and of Kongsberg in *Norway*.

DISTRIBUTION OF MERCURY.

IN THE UNITED STATES.—No mercury is known to have been found east
of the Mississippi River. It is produced in greater quantity than in any other
part of the world .from a mine at New Almaden, California, in one of the
side valleys of the San José.

IN FOREIGN COUNTRIES.—Nearly all the mercury supplied by Europe is
obtained from the mines of Almaden in *Spain*, and Idria in the *Austrian
Empire*. The mines of Almaden are situated in the province of La Mancha,
near the frontier of Estramadura. These mines have been worked longer
than any others in the world ; they were known to the Greeks at least 700
years before the Christian era. The present yield of the Spanish mercury
mines is about two and a half million of pounds per annum, or but little
short of that of the California mines. Mercury is obtained from several

localities in South America, but chiefly from the mines of Huancavelica, in *Peru*, which yield about 200,000 pounds per annum.

DISTRIBUTION OF TIN.

IN THE UNITED STATES.—Tin, which is used in the manufacture of tin plates, is nowhere obtained in the United States in any great amount. The only locality in the eastern part of this country where this ore has been found in any considerable quantity, is at Jackson, in Carroll County, New Hampshire, but it has not been obtained there to such extent as to render its manufacture profitable.

IN FOREIGN COUNTRIES.—The most productive tin mining region in the world is in Cornwall, *England*. The Cornish mines have been worked from a very early period, the metal from which formed an article of traffic with the Phœnicians and Greeks before the time of our Saviour. It is estimated that about 7,000 tons of tin are annually made from the ores of these mines. The most valuable tin mines on the continent of Europe are those of Erzge. birge, which are partly in *Saxony* and partly in *Bohemia*. One of the richest deposits of tin known is in the province of Tenasserim, in the *Malayan* peninsula. The best quality of tin is obtained from the island of *Banca*, at the extremity of the Malacca peninsula.

DISTRIBUTION OF COPPER.

IN THE UNITED STATES.—Copper is, next to gold and iron, the most important metal in the United States. The richest copper mining district in this country is that of Lake Superior. The occurrence of native copper on this lake was known to the Jesuit Fathers, who, in the latter half of the seventeenth century, traveled extensively in that region. The mines were not extensively worked until after the year 1844, when the country was ceded to the United States by the Chippewa Indians, and opened to settlement. Since that time, numerous companies have been organized, and large amounts of stock contributed, for the purpose of opening and working mines in this region. The principal mines are those of Kewenaw Point, Isle Royale, Ontonagon River, and Portage Lake. The Cliff Mine of Kewenaw Point has been worked for the longest period, and has yielded the greatest amount of metal.

Copper deposits are found at numerous localities in the valley of the Mississippi,—those in the neighborhood of Mineral Point, Wisconsin, being the best known. They are found also in the region which extends along the western slope of the Appalachian chain of mountains from New Hampshire to Georgia. The principal localities in this district occur at or near the following places : Warren, in *New Hampshire*; Orange, in *Vermont*; Bristol, Manchester, Litchfield, and Plymouth, in *Connecticut*; Belleville, Griggstown, Brunswick, Woodbridge, Greenbrook, Somerville, and Flemington, in *New Jersey*; Montgomery and Chester counties, in *Pennsylvania*; Liberty and New London, in *Maryland*; Fauquier County, in *Virginia*; Greensboro, in *North Carolina*; and Polk County, in *Tennessee*.

IN FOREIGN COUNTRIES.—The most noted copper mines of foreign countries are those of *Chile*, in South America, where copper is the most important product. The principal mines are located at Carrisal, north of the valley of Huasco; and at San Juan and La Higuera, between Huasco and Coquimbo. Numerous others are worked in the vicinity of Copiapo. The annual yield of copper in Chile is over 20,000 tons. The copper mines of Cornwall and Devonshire, in England, are highly celebrated; and have been worked longer, and have produced more of this metal than any other mines in the world. They are supposed to yield about 14,000 tons annually.

The other principal copper mining districts are those of the Ural Mountains and the Caucasus, in the *Russian Empire*; Mansfield, in *Prussia*; Upper Hungary, in *Austria*; and Namaqua Land, in *Southeastern Africa*. Copper is also obtained in *Norway*, *Sweden*, *Spain*, and in the *East Indies*, *Japan*, *Australia*, and *Cuba*.

DISTRIBUTION OF ZINC.

IN THE UNITED STATES.—The ores of zinc are distributed over the United States in great abundance. Some of the most important localities are at or near the following places: Easton, in *New Hampshire*; Wartsboro, in Sulli-

van County, *New York*; Sussex County, *New Jersey*; and Friedensville, in Lehigh County, *Pennsylvania*. Of the above mines, those in New Jersey and Pennsylvania have, as yet, yielded by far the greatest amount of this metal. The ores of zinc are plentifully distributed through the lead mines of the Mississippi valley.

IN FOREIGN COUNTRIES.—Zinc is procured in Great Britain, Austria, and Poland ; but the countries particularly distinguished for extensive zinc mines are *Belgium* and *Prussia*, which, together, yield more than eight tenths of all that is manufactured in the world. The great Belgian zinc works are in the province of Liège. The principal zinc district of Prussia is in the province of Upper Silesia, which produces nine tenths of the whole amount of this metal furnished by Prussia.

DISTRIBUTION OF LEAD.

IN THE UNITED STATES.—The lead mines of this country are abundantly scattered over its surface, and have yielded a larger amount in value of this metal than of any other, with the exception of iron, gold, and copper. The most productive mines are those of the Upper Mississippi lead region ; but there are numerous localities in the Atlantic States where considerable amounts of lead have been obtained. The most noted of the lead districts in the latter division are those of Rossie, in St. Lawrence County, Ancram, in Columbia County, Northeast, in Dutchess County, *New York*; Middletown, in *Connecticut*; and Christie and Montgomery counties, in *Pennsylvania*.

The great lead deposits of the Mississippi valley are known as the Upper Mississippi and the Missouri mines. The first of these divisions comprehends the lead region lying in the southwestern portion of Wisconsin and including a small part of the adjacent States of Illinois and Iowa. The principal mining centers of this division are Galena, in *Illinois*; Mineral Point, in *Wisconsin*; and Dubuque, in *Iowa*. The Mississippi runs along the western edge of this tract, and the course of the Wisconsin River is nearly parallel with the northern line, and distant from it only a few miles. The mines of this division yield annually upward of 18,000 tons. The second division embraces the mines of the State of *Missouri*, lying principally south of the Missouri River. The principal mines are in Washington County, near Big River and Mineral Creek.

IN FOREIGN COUNTRIES.—Lead mines are distributed through most of the countries of Europe, but those of Great Britain, France, and Spain are by far the most productive. The most extensive lead mining districts of Great Britain are in Cardiganshire and Montgomeryshire in Wales, and at Alston Moor, where the three counties of Northumberland, Durham, and Cumberland come together.

DISTRIBUTION OF IRON.

IN THE UNITED STATES.—Iron is so very extensively distributed throughout the United States, that only a mere mention of the States in which it is most abundant can be given in this article. Rich deposits of this metal are found in almost, if not quite, every State in the Union ; but the States in which they were wrought to the greatest extent in 1860 were, in the order of their rank in this respect, Pennsylvania, New York, New Jersey, Maryland, Kentucky, New Jersey, Tennessee, and Missouri. Pennsylvania excels all other States in the manufacture of iron, having yielded in 1860 five eighths of all that was made in the Union. The number of tons of iron made in the United States, according to the census of 1860, was 884,474. Owing to the cheapness of foreign iron, and the extensive importation of it, the iron mines of this country are not worked so largely as they otherwise would be.

IN FOREIGN COUNTRIES.—Great Britain stands pre-eminent for the abundance of its iron and the extent of its manufacture, the annual produce of its mines being estimated at about 3,000,000 tons. Of the other foreign countries noted for the production of iron, are France, Belgium, the Austrian Empire, the Russian Empire, Sweden, and Prussia.

NOTE.—For the approximate amount of metals produced in various countries about the year 1860, see the following table:

PRODUCT OF MINES IN VARIOUS COUNTRIES NEAR 1860.

Country.	Coal, tons.	Iron, tons.	Copper, tons.	Lead, tons.	Zinc, tons.	Tin, tons.	Salt, tons.	Gold, lbs. Troy.	Silver, lbs. Troy.	Mercury, lbs. Troy.
AMERICA :										
Canada	—	27,000....	6,000....	—	.....	....	160,000....	200....	—	—
New Brunswick..................	20,000....	6,500....	—	—	.....	....	40,000....	50....	—	—
Nova Scotia...;................	265,000....	22,000....	—	—	.....	....	26,000....	1,000....	50....	—
United States.................16,000,000....	885,000....14,400....18,000....	6,000....	....	320,000....260,000....	52.000....	1,200,000
Mexico	—	48,000....	3,000....	—	.....	....	120,000....	15,500....2,400,000....	300,000

Country.	Coal, tons.	Iron, tons.	Copper, tons	Lead, tons.	Zinc, tons.	Tin, tons.	Salt, tons.	Gold, lbs. Troy.	Silver, lbs. Troy.	Mercury, lbs. Troy.
Central America	—	....	2 000	—	—	—	16,000	1,200	160,000	—
Cuba	...	...	8,000	—	—	—	—	—	—	—
Columbia	20,000	32,000	—	—	—	—	300,000	15,000	20,000	—
Brazil	—	38,000	—	—	—	—	100,000	8,000	1,200	—
Bolivia	—	16,000	1,800	—	—	200	—	1,600	128,000	—
Peru	50,000	5,000	2,000	—	—	{ 1,600 {	50,000	2,800	340,000	200,000
Chile	100,000	10,000	22,000	—	—	...	—	4,800	200,000	—
EUROPE:										
Great Britain	72,000,000	3,000,000	16,000	63,000	3,000	6.500	800,000	200	48,000	—
France	5,500,000	800,000	5,600	45,000	—	—	360,000	—	20,400	—
Spain	1,000,000	82,000	800	34,000	35	16	156,000	40	137,000	2,500,000
Belgium	8,300,000	365,000	—	2,100	23,000	—	—	—	11,000	—
Prussia	7,600,000	160,000	1,700	9,500	35,000	—	200,000	—	60,000	—
Austria	1,820,000	280,000	3,500	8,200	1,600	60	356,000	6 000	95,000	400,000
Germany	500,000	140,000	600	12,000	40	135	300,000	100	120,000	—
Sweden and Norway	—	188,000	3,100	600	100	—	100,000	50	28,000	—
Russian Empire	200,000	218,000	8,000	1,000	4,000	—	500,000	120,000	65,000	—
Italy	—	28,000	400	630	—	—	200,000	—	—	....
OTHER COUNTRIES:										
Southern Asia and E. India Islands	1,000,000	300,000	4,500	5,000	—	6,000	1,000,000	80,000	200,000	—
Australia	100,000	10,000	4,200	200	160	—	100 000	286,000	22,000	—
Africa	—	100,000	1,600	—	—	—	1,000,000	10,000	—	...

PRINCIPAL MOUNTAINS, AND THEIR ELEVATION,

NORTH AMERICA.

Rocky Mts., between 70° N. lat. and Lewis and Clark's Pass (47° 30' N. lat.), 3,000 to 10,000 feet; from the latter to South Pass (42° N. lat.), 10,000 to 14,000; thence to El Paso (about 32° N. lat.), 15,000 to 7,000.

Feet.
Mt. Brown 15,690
Mt. Hooker 15,700
Fremont's Peak 13,570
Long's Peak 12,560
Pike's Peak 14,500
Spanish Peaks 11,000
Mountain Peaks in the southern part of the Plateau of Anahuac:
Toluca 15,751
Popocatapetl (V.,° highest peak in N. America) 17,884
Istaccihuatl 15,705
Perote 18,415
Orizaba (V.) 17,373
Peaks on the Plateau of Central America:
Volcano of Agua 13,578
Cartago 11,480
Northwest Coast Range:
Mt. Ilaeman (V.) 12,066
Mt. St. Elias 16,756
Mt. Fairweather 14,708
Cascade Range:
Mt. St. Helens .... about 13,000
Mt. Shasta 14,390
Sierra Nevada .... probably from 10,000 to 12,000

NOTE —The heights, given below, of the principal peaks of the Appalachian Mountain System are according to the recent barometric measurements of Professor Guyot.

White Mts. (average of eight highest peaks) 5,601
Mt. Washington 6,288
Mt. Adams 5,794
Mt. Jefferson 5,714
Green Mountains:
Mansfield Mt (the Chin) 4,430
Killington Peak 4,221
Adirondac Mts. (average of ten highest peaks) .... ? 4,975
Mt. Marcy or Tehawus 5,379
Catskill Mts. (average) .... ? 3,000
Alleghany Mts.: average of Blue Ridge, in. Pa., 1,100 feet; in Md. and Va., near Harper's Ferry, 1,800; of highest peaks in central Va., 4,000; of the entire ridge in N. C., 3,200; in Ga., 1,800;

* Peaks marked thus (V.) are volcanoes.

Alleghanies Proper, near lat. 37¼°, 2,650; near lat. 36°, for a distance of 150 miles, 5,000; at their terminus in Ala., 1,200.

Feet.
Peaks of Otter (the highest) 3,993
White Top (southern border of Va.) 5,530
Black Mts. (average of eight highest peaks) 6,610
Black Dome, or Mitchell's High Peak, or Clingman's Peak 6,707
Balsam Cone 6,671
Black Brother or Sandoz Mt. 6,619
Cattail Peak 6,611
Smoky or Unaka Mts. (highest range of the Appalachian System) average about 6,300 ? feet.
Smoky Dome or Clingman's Mt. 6,660
Mt. Guyot 6,636
Mt. Leconte (central peak) 6,612
Mt. Buckley 6,599

ISLANDS OF AMERICA.

Sierra Maestra (Cuba, West Indies):
Parco de Tarquino 7,900
Le Gran Piedra 5,800
Blue Mts. (Jamaica, West Indies) 6,739
Cibao Mts. (Hayti,   "   ) 7,200
Sierra de Languilla (Porto Rico, West Indies) 3,678
Mt. Misery (V.) (St. Christopher,   "   ) 3,712
Souffriere (V.) (Guadaloupe,   "   ) 5,500
Solfatara (V.) (Dominica,   "   ) 6,075
Mt. Pellee (Martinique,   "   ) 4,430
Volcano (St. Lucie,   "   ) 4,000
Mt. Garoa (St. Vincent,   "   ) 5,007
Sarmiento (East Tierra del Fuego) 6,900

SOUTH AMERICA.

Silla de Caracas (Sierra Castanera, Venezuela) 8,632
Horqueta (Sierra Nevada de Santa Marta, New Granada) 19,184
Andes Mts., or Cordillera de los Andes: between 5° N. lat. and 2° S. lat., average from 11,000 to 12,000 feet; between 15° and 19° S. lat., about 15,000; southward of 42° S. lat., about 3,000.
Tolima (V.) (New Granada) 18,320
Purace (V.)   " 17,008
Cumbal (V.)   " 16,824
Cyambe (Ecuador) 19,584
Antisana   " 19,370
Pinchincha (V.) (Ecuador) 15,986
Cotopaxi (V.)   " 18,875
Tunguragua (V.)   " 15,960
Chimborazo   " 21,424
Sangay (V.)   " 16,827
La Viuda (Peru) 15,968

	Feet.
Vilcanota (Peru)...	17,525
Chuquibamba (Peru)..	21,000
Arequipa (V.)..	18,373
Nevada de Sorata (Bolivia)................................	21,250
Illimani "	21,181
Sahama (V.) (Bolivia and Peru)..........................	22,350
Gualatieri (V.) " " "	21,960
Potosi (Bolivia)...	16,150
Portezuela Come Caballo (Chile and Argentine Republic)......	14,521
Cordel de la Laguna " " " "	15,575
Aconcagua " " " "	22,301
Tupungato " " " "	22,450
Maypu " " " "	17,664
Villa Rica (V.) " " " "	16,000
Osorno (V.) (Chile and Patagonia)........................	7,550
Minchinmadom (V.) (Chile and Patagonia)	8,000
Yanteles (V.) (Patagonia).................................	8,080
Mt. Stokes "	6,400
Parime Mts. (Venezuela, Guiana, and Brazil), average from 4,000 to 5,000 feet.	
Duida (Venezuela)	8,467
Maravaca "	10,500
Roraima (Venezuela and B. Guiana)	7,450
Mts. of Brazil :	
Itambe (Serra Espinhaco)..................................	5,755
Itacolumi (Serra Mantigueira).............................	5,750
Morro dos Canados (Serra dos Orgaos)	4,476

EUROPE.

Scandinavian Mts. (Norway and Sweden) :	
Sulitelma	6,178
Snee-haeten...	8,120
Skagstol-tind..	8,101
Hardanger-field..	5,748
Vosges Mts. (France), average from 2,000 to 3,000 feet.	
Cevennes Mts. (France), average from 2,000 to 3,000 feet.	
Mt. d'Or, highest peak.................................	6,196
Pyrenees Mts. (France and Spain), average from 7,000 to 9,000 feet.	
Pic Néthou or Maladetta..................................	11,168
Mt. Perdu..	10,904
Cantabrian Mts. (Spain), average from 4,000 to 6,000 feet.	
Pena de Panaranda..	10,998
Iberian Mts. :	
Sierra de Oca..	5,450
Sierra Molina..	4,500
Castilian Mts. (Spain and Portugal), average from 4,000 to 5,000 feet.	
Sierra de Gredos (Spain).................................	10,551
Serra d'Estrella (Portugal)	7,524
Mts. of Toledo (Spain), average from 3,000 to 4,000 feet.	
Sierra de Guadaloupe.....................................	5,115
Serra Morena (Spain).......................	
Sierra Nevada (Spain), average from 6,000 to 9,000 feet.	
Mulhacen...	11,658
Veleta..	11,382
Harz Mts. (Hanover) :	
Brocken..	3,658
Schwarz Wald (Baden), average from 2,000 to 3,000 feet.	
Jura Mts. (France and Switzerland), average from 3,000 to 4,000 feet.	
Reculet...	5,959
The Alps (France, Italy, Switzerland, and Austria), western part, average from 8,000 to 9,000 feet.	
Monte Viso	12,582
Mont Olan	13,120
Mont Cenis	11,457
Mont Blanc	15,760
Great St. Bernard...	11,003
Monte Rosa	15,203
Jungfrau	13,710
Finster-aar-Horn	14,106

	Feet.
Mont St. Gothard	10,595
Ortler Spitz	12,818
Gross Glockner..	12,776
Apennines (Italy), average from 3,000 to 5,000 feet.	
Monte Cimone...	6,975
Monte Sybilla	7,212
Monte Corno or Gran Sasso..............................	10,154
Monte Vellino...	7,851
Monte Amaro..	9,113
Vesuvius (V.)...	3,947
Erz Gebirge (Saxony and Bohemia), average from 2,000 to 3,000 feet.	
Riesen Gebirge (Silesia and Bohemia), average from 3,000 to 4,000 ft.	
Bohmer Wald (Bohemia and Bavaria), average from 3,000 to 4,000 ft.	
Carpathian Mts. (northern and eastern portions), average from 5,000 to 6,000 feet.	
Lomnitz Peak (group of Mt. Tatra)	8,779
Dinaric Alps (Turkey), average from 4,000 to 5,000 feet.	
Dinari..	6,667
Mt. Kom..	9,575
Balkan Mts. (Turkey), average from 2,000 to 3,000 feet.	
Schar..	8,805
Orbelus...	9,000
Mt. Pindus Chain (Turkey and Greece), average from 5,000 to 6,000 feet.	
Olympus (modern name, Lacha)...........................	9,754
Guiona	8,239
Parnassus (modern name, Liakhura)	8,068
Caucasus Mts. (Caucasia and Georgia), average 8,000 to 10,000 feet.	
Elburs...	18,493
Kabek..	16,592
Ural Mts. (between Europe and Asia), average from 2,000 to 2,500 feet.	
Obdorsk..	5,118
Konjakofski	5,307
Tagnai...	3,592
Iremel...	5,075
Yaman...	5,400

ISLANDS OF EUROPE, ETC.

Hekla (V.) (Iceland)	5,095
Oraefa Jokull "	6,409
Mts. of Kerry (Ireland) :	
Brandon..	3,120
Carran Tual (highest of McGillacuddy's Reeks)	3,404
Wicklow Mts. (Ireland) :	
Lugnaquilla	3,039
Northern Highlands (Scotland) :	
Ben Wyvis..	3,720
Ben Atow...	4,000
Grampian Hills (Scotland) :	
Cairn Gorm	4,095
Ben McDhui..	4,305
Ben Nevis..	4,368
The Lowthers (Scotland)...................................	3,150
Cheviot Hills (between England and Scotland)...............	2,658
Cross Fell (Pennine Chain, England).......................	2,901
Scaw Fell (Cumbrian Mts., ")......................	3,166
Snowdon (Wales)...	3,571
Plynlimmon "	2,463
Gran Pico (Azores or Western Is.).........................	7,911
Monte Rotondo (Corsica)	9,069
Genargentu (Sardinia).....................................	5,276
Mt. Etna (V.) (Sicily).....................................	10,874
Mt. Ida (modern name, Psilorati, I. of Candia)............	7,674

ASIA.

Kamtchatka Mts., average from 10,000 to 15,000 feet.	
Shiveluteh (V.)...	10,591
Kliuchevsk (V.)...	16,512
Altai Mts. (Siberia, Mongolia, and Ili), average from 5,000 to 7,000 ft.	
Bielucha..	11,063

Feet.

Thian Shan (northern part of Chinese Empire), average from 6,000 to 8,000 feet.

Yalo.. 8,500

Petsha..above 10,000

Kuen Lun (Tibet, Ili, and China proper), average from 16,000 to 18,000 feet.

Himalaya Mts. (Ladak, Tibet, and Hindoostan), average from 15,000 to 18,000 feet.

Chumalari... 23,929

Kunchinginga.. 28,176

Everest (highest known peak on the globe)............. 29,002

Gossainthan... 24,740

Dhawalagiri.. 28,080

Jawahir.. 25,670

Bolor Mts. (Ili and Toorkistan), highest peaks average from 15,000 to 20,000 feet.

Hindoo Koosh Mts. (Toorkistan and Afghanistan):

Hindoo Koosh Peak................................... 20,280

Sufaid Koh... 16,000

Suleiman Mts. (Afghanistan and Hindoostan):

Suleiman Peak.. 12,150

Elbroos Mts. (Persia):

Demavend, height lately ascertained.................. 21,500

Zacros Mts. (Persia and Turkey), average from 6,000 to 9,000 feet.

Sheikjwa... 10,423

Taurus Mts. (Turkey), average from 4,000 to 5,000 feet.

Ararat... 17,323

Argæus (modern name, Arjish-dagh)................... 12,869

Olympus (" " Chehish-dagh)................ 9,000

Mt. Lebanon Chain (north of Palestine), average from 6,000 to 7,000 feet.

Hermon (modern name, Jebel-esh-Sheikh).............. 10,000

Mts. of Sinai (Arabia), highest peak.................. 9,300

Aravulei Mts. (Hindoostan)........................about 3,600

Vindhya Mts. " average from 2,500 to 3,000 feet.

Gauts Mts. " eastern chain, greatest height 3,000 feet; western chain, average from 3,000 to 4,500.

Taddiamdamata...................................... 5,690

ISLANDS OF ASIA.

Kini Balu (Borneo).................................. 13,698

Slamat (V.) (Java).................................. 11,300

Indrapura (V.) (Sumatra)............................ 12,500

Ophir (modern name, Pacaman, in Sumatra)........... 9,603

Pedrotallagalla (Ceylon)............................ 8,326

AFRICA.

Atlas Mts. (Barbary), average from 3,000 to 4,000 feet.

Miltsin.. 11,400

Hentet.. 15,000

Egyptian and Nubian Mts., average from 3,000 to 4,000 feet.

Mt. Agrib... 5,600

Abyssinian Mts. :

Abba Yaret.. 15,200

Mts. of the Moon (Zanguebar):

Mt. Kenia (V.), supposed............................ 20,000

Killmandjaro.. 20,000

Nieuveldt Mts. (South Africa):

Spits Kop.. 10,240

Table Mt.. 3,582

Cameroon Mts. (Guinea), highest peak............... 13,780

Mt. Atlantika.......................................from 9,000 to 10,000

Mts. of Kong (Soudan and Guinea), average from 3,000 to 4,000 feet.

Mt. Loma.. 5,000

Mts. of Central Africa:

Mt. Dogem.. 4,500

Mt. Baghzen..............................from 3,000 to 3,500

ISLANDS OF AFRICA.

Peak of Teneriffe (V.), (Canary Is.)................. 12,182

Fogo (Cape Verde Is.)............................... 9,157

Feet.

Santiago (Cape Verde Is.).............................. 7,400

Clarence Peak (Fernando Po)........................ 10,655

Ankaratra Mts. (Madagascar), average from 8,000 to 10,000 feet.

Piton de Neiges (Bourbon)........................... 10,855

AUSTRALASIA, POLYNESIA, ETC.

Warragong Mts. or Australian Alps (New South Wales and Victoria), average 2,500 feet.

Mt. Wellington...................................... 6,510

Omeo.. 3,700

Mt. William... 4,000

Mt. Bryan (South Australia).......................... 4,500

Mt. Brown " 3,000

Gawler Range " average...................... 2,000

Darling Range (West Australia), average.............. 2,500

Toolbrunup.. 5,000

Mts. of Tasmania, principal chain; average............ 3,750

Mt. Humboldt....................................... 5,520

Mts. of New Ulster (New Zealand):

Mt. Edgecombe (V.)..............................about 10,000

Tongariro (V.)....................................... 6,200

Ruapaho.. 9,000

Egmont (V.)... 8,889

Mts. of New Munster (New Zealand):

Kaikora... 9,900

Mt. Arthur.. 8,000

Rowley Peak... 5,000

Opureone (Tahiti, Society Is.)....................... 8,000

Mauna Kea (V.) (Hawaii, Sandwich Is.).............. 18,587

Mauna Loa " " 13,175

Mt Erebus " (South Victoria Land)........... 12,400

PRINCIPAL RIVERS OF THE WORLD,

SHOWING THEIR SOURCES—THE COUNTRIES THROUGH WHICH THEY FLOW —THE SEAS INTO WHICH THEY FALL—THEIR APPROXIMATE LENGTH AND NAVIGABLE DISTANCE IN ENGLISH MILES.

After the name of each river follow—1st (in parenthesis), that of the district in which it rises; 2d, *in Italics*, the countries through which it flows; 3d, the sea or ocean into which it falls; and, lastly, its length, inclusive of the principal windings of its stream, and its navigable distance, when it can be ascertained.

NORTH AMERICA.

	Total Length.	Nav'e Dist.
MACKENZIE, including Peace R. (Rocky Mts.).—*British Possessions:* Arctic Ocean.................................. 2,000		
NELSON, including Saskatchewan R. (Rocky Mts.).—*British Possessions:* Hudson Bay................................. 1,400		
ST. LAWRENCE, with great lakes (Plateau of Minnesota).—*United States, Canada:* Gulf of St. Lawrence 2,050		

Navigable for ships 550, and by artificial means for smaller vessels to within 150, miles of its source.. 1,900

OTTAWA (table-land of Canada).—*Canada:* St. Lawrence River...........	450	100
ST. MAURICE " " "	300	100
SAGUENAY " " "	230	70
SORELLE, or St. John, with Lake Champlain "	175	175
ST. FRANCIS (Lake St. Francis)—*Canada:* "	130	
CHAUDIÈRE (Lake Megantic), " "	125	4
ST. JOHN (divide between the U. States and Canada).—*United States, Canada, and New Brunswick:* Bay of Fundy.	330	80
PENOBSCOT " " —*Maine:* Atlantic Ocean........	275	50
KENNEBEC " " " "	200	42
ANDROSCOGGIN " " " "	140	18
CONNECTICUT " " — *Vermont, N. Hampshire, Massachusetts, Connecticut:* Long Island Sound..........	410	50
MERRIMAC (White Mts.).—*N. Hampshire, Mass.:* Atlantic Ocean............	170	5
HOUSATONIC (Taghkanic Mts.).—*Mass., Conn.:* Long Island Sound.........	130	14
HUDSON (Adirondac Mts.).—*N. York:* Atlantic Ocean.	310	160
MOHAWK " " Hudson River....................	135	
GENESEE (Alleghany Mts.) " Lake Ontario...................	145	6
DELAWARE (Catskill Mts.).—*N. York, N. Jersey, Penn.:* Delaware Bay......	320	75
SUSQUEHANNA (Alleghany Mts.).—*N. York, Penn., Md.:* Chesapeake Bay....	450	
POTOMAC " —*Md., Va.:* " ...	380	115

	Total Length	Nav's Dist.
JAMES (Alleghany Mts.),—*Virginia:* Chesapeake-Bay	875	110
ROANOKE " —*Virginia, N. Carolina:* Albemarle Sound	275	75
NEUSE " —*North Carolina:* Pamlico Sound	250	45
CAPE FEAR " —*North Carolina:* Atlantic Ocean	250	100
GREAT PEDEE " —*N. and S. Carolina:* "	375	14
SANTEE " "	860	150
SAVANNAH " —*Georgia, South Carolina:* Atlantic Ocean	830	185
ALTAMAHA " —*Georgia:* Atlantic Ocean	820	190
ST. JOHN'S (Everglades),—*Florida:* "	230	175
APPALACHICOLA, with Chattahoochee River (Alleghany Mts.),—*Georgia, Alabama, Florida:* Gulf of Mexico	470	270
FLINT (Alleghany Mts.),—*Georgia:* Appalachicola River	275	80
MOBILE, with Alabama and Coosa Rivers (Alleghany Mts.),—*Georgia, Alabama:* Gulf of Mexico	1,090	887
TOMBIGBEE (Alleghany Mts.),—*Mississippi, Alabama:* Gulf of Mexico	475	880
BLACK WARRIOR " —*Alabama:* Tombigbee River	300	183
PEARL (table-land, Miss.),—*Mississippi:* Gulf of Mexico,	380	
MISSISSIPPI (Lake Itasca),—*Min., Wis., Iowa, Ill., Mo., Ky., Tenn., Ark., Miss., La.:* Gulf of Mexico	2,500	2,200
RED (Llano Estacado),—*Tennessee, Arkansas, La.:* Mississippi River	1,200	375
WASHITA (Ozark Hills),—*Arkansas, Louisiana:* Red River	460	250
YAZOO (table-land, Miss.),—*Mississippi:* Mississippi River	350	65
ARKANSAS (Rocky Mts.),—*Kansas, Indian Ter., Ark.:* Mississippi River	2,000	645
WHITE (Ozark Ridge),—*Missouri, Arkansas:* Arkansas River	550	140
ST. FRANCIS " " Mississippi River	825	100
OHIO, with Alleghany River (Alleghany Mts.),—*Pennsylvania, New York, Ohio, Virginia, Kentucky, Indiana, Illinois:* Mississippi River	1,200	948
MONONGAHELA (Alleghany Mts.),—*Virginia, Pennsylvania:* Ohio River	300	140
MUSKINGUM (table-land, Ohio),—*Ohio:* "	250	100
GREAT KANAWHA (Alleghany Mts.),—*Virginia* "	400	66
SCIOTO (table-land, Ohio),—*Ohio:* "	200	130
LICKING (Cumberland Mts.),—*Kentucky:* "	180	70
MIAMI (table-land, Ohio),—*Ohio:* "	100	75
KENTUCKY (Cumberland Mts.),—*Kentucky:* "	300	60
GREEN " " —*Indiana, Illinois:* "	500	300
CUMBERLAND (Cumberland Mts.),—*Tennessee, Kentucky:* "	550	120
TENNESSEE, with Holston River (Alleghany Mts.),—*Virginia, Tennessee, Alabama, Mississippi, Kentucky:* Ohio River	1,200	276
KASKASKIA (Grand Prairie),—*Illinois:* Mississippi River	300	7
MISSOURI (Rocky Mts.),—*Nebr., Min., Iu., Kan., Mo.:* Mississippi River	3,100	1,806
From the mouth of the Missouri to the Gulf of Mexico the distance is 1,250 miles—making the total length of the Missouri 4,450 miles.		
YELLOW-STONE (Rocky Mts.),—*Nebraska:* Missouri River	750	250
PLATTE " "	1,200	96
KANSAS (Great Plains),—*Kansas:* "	600	150
OSAGE " —*Kansas, Missouri:* "	450	185
ILLINOIS, with Kankakee River (Indiana),—*Ind., Ill.:* Mississippi River	485	230
DES MOINES (Coteau des Prairies),—*Minnesota, Iowa:* "	400	200
IOWA " —*Iowa:* "	275	170
ROCK (Wisconsin),—*Wisconsin, Illinois:* Mississippi River	400	120
WISCONSIN (Onionagon Ridge),—*Wisconsin:* "	260	70
CHIPPEWA " "	225	30
ST. CROIX " —*Wis., Min.:* "	300	100
ST. PETERS (Coteau des Prairies),—*Min.:* "	460	
SABINE (Red River divide),—*Texas:* Gulf of Mexico	580	60
TRINITY " " "	700	250
BRAZOS " " "	690	220
COLORADO (Llano Estacado), " "	850	100
NUECES (Guadelupe Mts.),—*N. Mexico, Texas:* Gulf of Mexico	2,000	700
RIO GRANDE (Rocky Mts.),—*N. Mexico, Texas, Mexico:* Gulf of Mexico	100	20
COATZACOALCOS (Cordilleras),—*Mexico:* "	250	120
USUMASINTA " —*Guatemala, Mexico:* "		
HUMUYA, including Ulua (Cordilleras of C. Amer.),—*Honduras:* Caribbean S.	250	
CHOLUTECA, " " G. of Fonseca.	150	
SAN JUAN (Cordilleras of C. Amer.),—*L. Nicaragua, Nicaragua:* Carib. S.	220	220
MEXCALA, with the Zacatula (Cordilleras),—*Mexico:* Pacific Ocean	600	
RIO GRANDE DE SANTIAGO (Lake Lerma),—*Mex., Quera., Micho., Guan., Guadalaxara:* Pacific Ocean		
COLORADO (Rocky Mts.),—*Oregon, Utah, N. Mex., California, Mexico:* Gulf of California	1,850	250
GILA (Sierra Madre),—*New Mexico:* Colorado River	650	
SAN JOAQUIN (Sierra Nevada),—*California:* Sacramento River	350	50
SACRAMENTO " " San Francisco Bay	450	150
COLUMBIA (Rocky Mts.),—*British Possessions, Washington, Oregon:* Pacific Ocean	1,800	185
LEWIS' FORK (Rocky Mts.),—*Oregon, Washington:* Columbia River	750	
CLARK'S FORK " —*Washington:* "	500	
WILLAMETTE (Cascade Mts.),—*Oregon:* "	225	15
FRASER (Lake Stuart),—*British Possessions:* Gulf of Georgia	1,000	
KWICHPAK (Rocky Mts.), " *Russ. Possessions:* Behring Sea.	1,000	

SOUTH AMERICA.

	Total Length	Nav's Dist.
MAGDALENA (Andes),—*New Granada:* Caribbean Sea	860	400
ORINOCO (Parime Mountains),—*Venezuela:* Atlantic Ocean	1,200	800
ESSEQUIBO (Sierra Acaray),—*British Guiana:* "	450	60
DEMERARA " —*Guiana:* Atlantic Ocean	180	
BERBICE " " "	250	165
CORENTYN " " "	375	150
SURINAM " " "	300	150
AMAZON, or Maranon (Andes),—*Ecuador, Brazil:* Atlantic Ocean	3,900	3,662
NEGRO (Plateau of New Granada),—*N. Granada, Brazil:* Amazon Riven	1,400	300
UCAYALI (Andes of Peru),—*Peru:* "	1,000	200
MADEIRA (Andes of Bolivia),—*Bolivia, Brazil:* "	2,000	1,000
TOCANTINS, with Peru (Serra dos Vertentes),—*Brazil:* Atlantic Ocean	1,400	500
PARNAHIBA (Serra Borborema), "	650	
SAN FRANCISCO (Mountains of Brazil), "	1,500	
PARANA, with Rio de la Plata (Serra dos Vertentes),—*Brazil, Paraguay, Argentine Republic:* Atlantic Ocean	2,350	1,250
PARAGUAY (Geral Mts.),—*Brazil, Argentine Republic:* Parana River	1,500	1,000
PILCOMAYO (Andes of Bolivia),—*Bolivia, Argentine Repub.:* Paraguay R.	906	
VERMEJO (Plateau of Despoblado, "	750	
SALADO " " Parana R.	1,000	600
URUGUAY (Serra de Mar),—*Brazil, Uruguay:* Rio de la Plata	800	

EUROPE.

	Total Length	Nav's Dist.
VOLGA (Plateau of Valdai),—*Russia:* Caspian Sea	2,200	1,800
DON (Great European Plain), " Sea of Azof	1,000	700
DNIEPER " " Black Sea	1,200	1,100
DNIESTER (Carpathian Mts.),—*Galizia, Russia:* Black Sea	500	48
DANUBE (Schwarz Wald),—*Germany, Hungary, Turkey:* Black Sea	1,680	1,500
PO (Alps),—*Italy:* Adriatic Sea	450	280
TIBER (Apennines),—*Italy:* Mediterranean Sea	210	47
RHONE (Alps),—*Switzerland, France:* Mediterranean Sea	490	200
EBRO (Cantabrian Mts.),—*Spain:* "	420	290
GUADALQUIVER (table-land of Spain),—*Spain:* Atlantic Ocean	290	150
GUADIANA " —*Spain, Portugal:* Atlantic Ocean	510	90
TAGUS " " "	510	90
DOURO " " "	460	110
GARONNE (Pyrenees),—*France:* Bay of Biscay	350	260
LOIRE (Cevennes), " "	570	840
SEINE (Plateau of Langres),—*France:* English Channel	480	850
SCHELDT (Plain of North France),—*France, Belgium:* North Sea	250	140
RHINE (Alps),—*Switzerland, Germany, etc.:* "	760	550
WESER (Hatz Mts.),—*Germany:* "	330	25
ELBE (Bohmer Wald Mts.), "	690	480
ODER (Carpathian Mts.), " Baltic Sea	550	810
VISTULA (Carpathian Mts.),—*Poland, Prussia:* Baltic Sea	628	
NIEMEN (Great European Plain),—*Russia, Prussia:* Baltic Sea	400	
DUNA (Plates of Valdai),—*Russia:* Gulf of Riga	550	
NEVA (from Lake Ladoga), " Gulf of Finland	45	
GOTA (from Lake Wener),—*Sweden:* Cattegat	58	
GLOMMEN (Scandinavian Mts.),—*Norway:* Skager-rack	400	
DWINA (Great European Plain),—*Russia:* White Sea	750	
PETCHORA (Ural Mts.), " Arctic Ocean	900	
SPEY (Grampian Mts.),—*Scotland:* North Sea	96	
TAY " "	120	85
FORTH (Grampian Mts.), "	150	40
TWEED (Lowther Hills), "	96	10
TYNE (Pennine Chain),—*England:* "	70	12
HUMBER { OUSE (Pennine Chain) / TRENT (Staffordsh. Moorl'ds) } *England:* North Sea	180	40
THAMES (Cotswold Hills), "	215	170
SEVERN (Welsh Mts.),—*Wales, England:* Bristol Channel	240	150
DEE " " Irish Sea	75	20
MERSEY (Pennine Chain),—*England:* Irish Sea	170	40
EDEN " " Solway Frith	80	15
CLYDE (Lowther Hills),—*Scotland:* Firth of Clyde	98	85
LIFFEY (Mts. of Wicklow),—*Ireland:* Irish Sea	75	
BARROW (Slieve Bloom Mts), " Waterford Harbor	105	80
SUIR " "	80	40
SHANNON (Mts. of Cavan), " Atlantic Ocean	224	200
BOYNE (Central Plain),—*Ireland:* Irish Sea	80	19

ASIA.

	Total Length	Nav's Dist.
OBI (Altai Mts.),—*Siberia:* Arctic Ocean	2,550	900
YENESEI " "	2,900	540
LENA " "	2,400	800
AMOOR (Mongolia),—*Mongolia, Manchooria:* Gulf of Tartary	2,800	
HOANG-HO (Kuenlun Mts.),—*China:* Yellow Sea	2,600	500
YANG-TSE-KIANG " "	3,200	700
CHOO-KIANG (Nan-ling Mts.),—*China:* China Sea	1,100	370
MAY-KIANG (Tibet),—*Tibet, China, Cambodia:* China Sea	1,800	500
MEINAM (Mt. of Yunnan),—*China, Laos, Siam:* Gulf of Siam	900	
SALUEN (Plateau of Tibet),—*Tibet, China, etc.:* Gulf of Martaban	1,200	
IRAWADY " —*Birmah:* Bay of Bengal	1,100	600
BRAMAHPOOTRA " —*Assam, Hindostan:* Bay of Bengal	1,500	650
GANGES (Himalaya Mts.),—*India:* Bay of Bengal	1,460	900

	Total Length.	Nav'e Dist.
GODAVERY (The Ghauts),—*Hindoostan :* Indian Ocean	70	60
KISTNAH " " "	650	
CAUVERY " " "	470	
INDUS (Plateau of Tibet),—*India :* Arabian Sea	1,700	943
EUPHRATES (Plateau of Armenia),—*Turkey in Asia :* Persian Gulf	1,700	1,200
TIGRIS " " Euphrates River	1,150	820
KUR "—*Georgia :* Caspian Sea	520	
JORDAN (Mts. of Lebanon),—*Palestine :* Dead Sea	287	
URAL, or Ialk (Ural Mts.),—*between Europe and Asia :* Caspian Sea	1,150	
AMOO, or Jihon (Plateau of Pamir),—*Toorkistan :* Sea of Aral	1,800	

AFRICA.

NILE (Plateau Central Africa),—*Donga, Nubia, Egypt :* Mediterranean Sea	3,600	
SENEGAL (Mts. of Western Soudan),—*Senegambia :* Atlantic Ocean	900	400
GAMBIA " " "	650	300
QUORRA, or Niger (Mts. of Kong),—*Soudan :* Gulf of Guinea	2,500	700
ZAIRE, or Congo (Plateau Central Africa),—*Congo :* Atlantic Ocean	1,200	
COANZA "—*Angola :* "	500	
GARIEP, or Orange (Nieuveldt Mts.),—*S. Africa :* "	1,050	
ZAMBESE (interior of Africa),—*East Africa :* Indian Ocean	1,500	

OCEANICA.

MURRAY (Warragong Mountains),—*Australia :* Indian Ocean	1,280	875
SWAN RIVER (Darling Range), " "	250	10
DERWENT (Lake St. Clair),—*Tasmania :* Pacific Ocean	140	26
WAIKATO,—*New Zealand :* Pacific Ocean.		

TABLE* OF MEAN TEMPERATURE OF EACH SEASON AND
OF THE WHOLE YEAR.

COUNTRIES.	Lat. N.	Long. W.	Elevation above lev. of sea.	Winter.	Spring.	Summer.	Autumn.	Year.
	° '	° '	Feet.	°	°	°	°	°
Melville Island	74 47	110 48		−29.45	−3.19	37.08	−0.48	1.24
Port Bowen	73 14	88 56	..	−25.09	−5.77	34.40	10.58	3.58
Boothia Felix	69 59	92 1	..	−21.71	−5.21	38.04	9.69	8.70
Winter Island	66 11	83 11	..	−20.47	6.85	31.80	17.58	8.92
Fort Simpson	62 11	121 82	250	−11.04	26.10	59.16	26.24	25.12
Iakutsk	62 1	−129 44	..	−36.87	15.61	61.72	12.76	18.43
Fort Franklin	65 12	123 13	230	−16.66	14.05	50.28	21.12	17.18
Nain	57 10	61 50	..	8.66	24.57	47.9¹	35.16	27.89
Tornea	65 24	−23 47	..	6.41	37.88	57.89	32.10	31.06
Irkutsk	52 17	−104 17	1855	0.90	35.89	61.50	32.77	32.69
Kazan	55 45	−49 7	150	6.34	38.90	62.89	36.91	35.45
Umea	63 50	−20 16	..	13.47	38.16	57.42	37.45	35.87
St. Petersburg	59 56	−30 18	..	18.66	37.06	61.68	41.02	39.61
Moscow	55 45	−37 36	400	15.90	40.98	68.97	39.94	40.02
Quebec	46 48	71 17	..	14.15	39.05	68.08	46.10	41.85
Christiana	59 55	−10 45	74	23.18	40.08	59 88	42.69	41.45
Stockholm	59 21	−18 4	128	26.04	38.21	60.43	44.41	42.27
Halifax	44 89	68 88	..	21.00	31.67	61.00	44.67	40.08
Montreal	45 31	73 85	..	17.79	45.76	71.40	48.05	45.76
Toronto	43 40	79 22	340	25.43	42.84	64.63	44.81	44.81
Warsaw	52 13	−21 1	851	24.91	48.05	68.91	45.41	44.15

* In the above Table, a *minus* sign (−) placed before a degree of lat. indicates that the
lat. is *south*, and placed before a degree of long. denotes that it is *east* long.

COUNTRIES.	Lat. N.	Long. W.	Elevation above lev. of sea.	Winter.	Spring.	Summer.	Autumn.	Year.
	° '	° '	Feet.	°	°	°	°	°
Delaware	42 16	74 58	1884	26.64	43 43	67.28	46.03	46.88
Leipsic	51 20	−12 22		31.79	45.72	60.23	47.86	46.41
Copenhagen	55 41	−12 35		31.31	43.54	62.70	48.70	46.56
Breslau	51 07	−17 02	878	30.19	45.78	68.61	48.42	46.74
Danzic	54 20	−18 41		30.01	43.57	61.92	47.06	45.64
Northumberland				37.08	44.64	57 87	47.64	46.68
Edinburgh	55 56	8 11	220	38.45	45.02	57.17	47.89	47.13
Meiningen	50 40	−10 80		32.61	46.58	63.46	48.90	47.89
Bremen	58 05	−8 49		38.10	47.11	63.30	48.99	48.13
Berlin	58 90	−18 24	100	31.45	47.41	64.06	49.28	48.16
Boston	42 21	71 4		28.99	46.09	69.04	50.46	48.47
Dublin	53 21	6 11		39.83	47.16	59.57	49.08	49.05
Plymouth	50 23	4 7		44.88	49.68	60.87	52 91	52.08
Geneva	44 19	−6 10	1258	34.04	52.21	70.36	54.2⅛	52.71
Vienna	48 13	−16 23	450	31.95	51.50	69.40	51.16	51.08
Brussels	50 51	−4 22	202	38.01	49.04	64.04	51.00	50.68
London	51 30	0 5		39.50	49.06	62.93	51.53	50.88
Neufchatel	46 59	6 55	1850	35.68	50.08	66.50	52 07	51.08
Paris	48 50	−2 2⁴	114	37.85	50.69	64.56	52.20	51.31
Bristol	51 27	2 86		40.83	50.33	64.33	51.67	51.67
Auckland	86 51	−174 54		66.92	58.88	50.75	55.68	58.56
Hobart Town	42 58	−147 98		63.06	61.56	42.14	53.75	52.87
Turin	45 11	−7 41	807	33.46	58.75	71.51	53.50	58.13
Philadelphia	89 67	75 1		30.07	49.80	71.86	51.87	50.78
New York	40 43	74 1		30.12	52.06	70.93	52.99	51.56
Pekin	39 54	−116 20		25.23	55.51	75.17	54.99	53.28
Washington	38 57	76 55	115	37.76	56.19	70.74	56.87	56.89
Nantes	47 13	1 88		40.63	54.65	68.68	55.63	54.90
Trieste	45 38	−18 46		39.44	58.69	71.89	56.69	55.30
Bordeaux	44 50	0 85		43.10	56.08	71.06	57.58	57.03
Madrid	40 25	3 42	1989	43.16	55.64	76.40	57.44	58.16
Marseilles	43 18	−5 22	140	45.92	55.91	73.98	59.21	58.82
Rome	41 54	−12 96	160	46.72	58.86	74.24	62.75	60.49
Montevideo	34 54	56 13		71.88	68.00	57.83	64.77	65.88
Quito	−0 14	78 45	8970	77.60	60.11	59.71	60.61	73.81
Naples	40 52	−14 15		47.85	57.56	74.88	61.46	60.26
Genoa	44 91	−8 54		46.98	58.51	73.06	62.97	61.05
Lisbon	38 42	9 9		52.59	59.66	70.94	62.48	61.40
St. Helena	−16 55	5 48	1764	68.87	64.90	59.41	55.88	61.40
Mexico	19 25	99 6	6990	58.64	68 48	65.28	62.10	60.60
Norfolk (Virg'n, U.S.)	36 58	76 16		46.95	62.91	78.80	66.69	63.71
Messina	88 11	−15 84		51.97	61.48	77.14	69 91	65.57
Cape of Good Hope	−84 11	18 26		66.95	62.00	54.29	60.79	60.77
Alexandria	31 13	−29 48		54.43	69.81	78.89	69.04	68.09
Barcelona	41 28	−6 12	204	50.18	60.37	77.00	64.51	65.08
Funchal	82 84	16 56	80	63.50	64.46	71.00	70.68	67.61
Constantinople	41 0	−29 0		40.94	53.00	71.38	60.56	56.47
Cadiz	36 ?3	6 18		52.90	58.68	70.48	65.85	62.06
Jerusalem	31 47	35 14	2500	49.81	60.59	78.88	65.55	62.85
Ionian Islands	38 0	−21 0		58.06	60.37	76.59	69.04	68.08
Malta	35 54	−14 84		58.06	60.37	67.00	60.64	63.08
New Orleans	29 58	90 7		55.80	72.08	82.04	69.28	69.80
Bermuda	82 90	64 50		58.76	68.74	75.30	71.90	67.40
Gibraltar	36 07	5 21		57.98	66.95	77.52	61.76	67.44
Tunis	36 48	10 11		55.76	64.54	88.00	71.86	68.77
Canton	23 8	−113 16		54.88	69.88	82.00	72.88	69.58
Cairo	30 2	−81 15		56.22	72.58	86.10	71.48	72.17
Bagdad	88 21	−44 29		49.62	75.04	96.18	77.14	73.74
Rio Janeiro	22 54	43 16		79.15	74.70	68.60	73.56	73.75
Mocha	13 20	−50 44		79.78	89.07	83.88	80.51	76.91
Havana	28 9	82 28		71.88	76.22	81.85	78.12	77.17
Vera Cruz	19 12	96 9		70.58	77.00	81.92	78.96	77.08
Seringapatam	12 45	−76 51	2412	71.88	82.07	74.97	74.67	75.76
Benares	25 18	−83 56	800	62.17	87.78	87.24	79.87	80.26
Aya	21 50	96 5		68.92	81.92	85.59	75.14	78.80
Calcutta	22 35	−89 20	80	72.25	87.80	88.72	83.00	82.41
Bombay	18 56	−78 54		77.44	85.25	82.94	81.54	81.27
Singapore	1 17	−108 50		78.51	81.05	81.61	80.82	80.63
Batavia	−6 9	−106 58		78.67	78.50	78.16	77.00	78.83
Trincomalee	8 84	81 22		77.58	79.93	88.90	81.85	80.75
Madras	18 4	−80 19		77.06	89.09	86.18	81.45	81.94

CPSIA information can be obtained
at www.ICGtesting.com
Printed in the USA
BVHW050016061118
532207BV00021B/2154/P